Textile Industry and Environment

Edited by Ayşegül Körlü

Published in London, United Kingdom

IntechOpen

Supporting open minds since 2005

Textile Industry and Environment
http://dx.doi.org/10.5772/intechopen.75336
Edited by Ayşegül Körlü

Contributors
Ayşegül Körlü, Pablo Colindres Bonilla, Fabien Salaün, Miray Emreol Gönlügür, Pinar Celik, Tuba Bedez Ute, Memik Bunyamin Uzumcu

First published in London, United Kingdom, 2019 by IntechOpen
IntechOpen is the global imprint of INTECHOPEN LIMITED, registered in England and Wales, registration number: 11086078, The Shard, 25th floor, 32 London Bridge Street
London, SE19SG - United Kingdom
Printed in Croatia

British Library Cataloguing-in-Publication Data
A catalogue record for this book is available from the British Library

Additional hard copies can be obtained from orders@intechopen.com

Textile Industry and Environment
Edited by Ayşegül Körlü
p. cm.
Print ISBN 978-1-83880-027-7
Online ISBN 978-1-83880-028-4

For EU product safety concerns:
IN TECH d.o.o., Prolaz Marije Krucifikse Kozulić 3, 51000 Rijeka, Croatia, info@intechopen.com or visit our website at intechopen.com.

Meet the editor

Prof. Dr. Ayşegül Ekmekci Körlü received her MSc and PhD degrees in Textile Engineering from Ege University. She has been employed at the Department of Textile Engineering, Engineering Faculty, Ege University, since 1988. Currently, her main research interests are sustainable and ecological finishing processes, pretreatment of natural fibers, functional textiles, and occupational safety in the textile industry. She has coauthored numerous publications and has taken part in the management of many national and international projects.

Contents

Preface

Environment and energy are two of the most important issues facing humanity today. The textile industry consumes a considerable amount of energy and water during the manufacturing processes. Energy and water consumption is closely associated with environmental issues, because production also means environmental pollution and water and energy shortage. Environmental problems caused by the textile industry can be studied under four classes: air, water, solid waste pollution, and noise. Waste management for solving environmental problems is realized by the utilization of cleaner technologies, reuse, material recycling, material and energy recovery, storage without recycling, or disposal by burning.

In this book, the relationship between the textile industry and the environment is examined from four different viewpoints, for which the basis is the waste management steps specified above. Recycling of spinning mill wastes, ozone usage that provides less chemical and water utilization, reuse of treated water in the dyeing processes, and approaches in the treatment of waste waters of dyeing plants and finishing factories are solutions offered to reduce environmental pollution arising from textile production processes. On the other hand, application of phase change materials to textile materials is examined in one chapter. The possibility of utilizing textile materials to which phase change materials are applied, not only for comfort purposes but also as energy storage materials, means that technical textiles could be a solution for energy storage. Briefly, the relationship between the textile industry and the environment both as a source of and as a solution to pollution is mentioned in this book.

I thank all the authors who have contributed to this book and hope that it will be helpful to the readers.

Dr. Ayşegül Körlü
Professor,
Ege University, Engineering Faculty,
Department of Textile Engineering,
İzmir, Turkey

Chapter 1

Sustainable Production Methods in Textile Industry

Miray Emreol Gönlügür

Abstract

The textile industry is part of the industries that continuously harm the environment because of the high water consumption and the presence of various pollutants in the wastewater. Wastewater treatment is lacking or includes only physical treatment in underdeveloped and developing countries due to installation and operating costs of a treatment plant. As a result, a broad spectrum of hazardous and toxic substances, such as (azo) dyes, heavy metals, acids, soda, and aromatic hydrocarbons, pollute precious sources of clean water, in which untreated water is discharged. The main solution to this problem is to reduce the treatment cost. For this purpose, the process should be optimized to reduce the amount of water and chemicals. In this chapter, first studies on the reference document (BAT) referred by the European Council are reviewed. Minimizing production costs, obtaining high-quality products, and reducing the amount and the pollutant content of wastewater are complex problems that cannot be solved by the conventional optimization methods. Therefore, nonconventional optimization methods applied on the textile processes are also reviewed from the latest studies in the literature.

Keywords: sustainability, BAT, optimization, metaheuristics, genetic algorithm

1. Introduction

Textile industry is the foremost sector in terms of the discharged volume and the composition of the wastewater [1]. Most of the textile production processes, such as scouring, washing, dyeing, bleaching, sizing, and finishing, consume large volumes of fresh water and discharge large volumes of effluent which are generally with intense color, high concentration of organic compounds, and large variations in composition [2, 3]. Especially wet process, which has five main stages including pretreatment, dyeing, finishing, drying, and quality control, is the major part of the textile industry due to the long processing time and technical complexity [4]. Specific water consumption range is given as 10–645 L/kg product for the textile industry and 21–645 L/kg for the mills with finishing and dyeing processes in (ref BAT-EC). In another source, it is reported that the consumed amount of water could reach to 932 L/kg product depending on the fiber and applied technology [5–7].

In recent years, depleted resources, global warming, and climate change resulting in challenging regulations enforce manufacturing enterprises to give efforts to reduce the waste of the processes, which motivate firms for the sustainable (cleaner) production. Besides environmental considerations, the effective planning of the production is also obligatory to reduce the production costs to be able to compete with other manufacturers. Moreover, an empirical study built on the theoretical model derived from the Cobbe-Douglas production function revealed that customers prefer the environmentally friendly goods produced by using cleaner production principles in Japan [8].

Cleaner production is defined in the report of the United Nations Environment Program (UNEP) [9] as "the continuous application of an integrated preventive environmental strategy to processes, products, and services, to increase overall efficiency, and reduce risks to humans and the environment." Unlike the end-of-pipe (EOP) treatment, which takes the design of the production fixed and attempts to solve the problem after the occurrence of pollution, cleaner production approach aims to solve the problem before it happens by considering pollution and wastes mostly as a result of the inadequacy, inefficiency, and ineffectiveness of the design, utilization of resources, and production processes stages. Necessary developments in these stages are suggested to provide sustainability to the processes [10].

The European Commission referred the best available techniques (BAT) reference document for the textile industry in 2003 [11]. The purpose of the document is to achieve a high level of protection for the environment as a whole. The document includes general information on the industrial sector concerned and on the industrial processes used within the sector, data and information about emission and consumption levels, emission reduction and other techniques that are considered to be most relevant for determining BAT and BAT-based permit conditions, and the techniques and the emission and consumption levels that are considered to be compatible with BAT.

Optimization ensures the most efficient use of existing processes. In the production facilities, the problem of reducing the use of water and energy used to produce high-quality products for a limited time is a complex problem to solve with conventional optimization methods. The consumption of energy and by means of noncostly changes and production planning, the use of waste treatment, and the reduction of water usage in the process, the environment will be covered. Multi-objective optimization methods target this type of complex problems.

Metaheuristic optimization methods are problem-independent techniques unlike the heuristic methods. However, it is necessary to introduce the intrinsic parameters of the problem to adapt the technique. Application of metaheuristic optimization methods to textile processes is a novel subject, and there are different perspectives on it. In some of the studies, the lot size and scheduling [12] are tried to optimize, while some of them try to model the end-of-pipe treatments [13–19]. However, only a few studies are meant to reduce the amount of wastewater and improve the process for clean production [4, 20–23].

In a recent review [24], different approaches and methods in water conservation for textile wet processing industry were effectively classified into five main groups that have many subgroups: (i) water conservation through textile wastewater treatment and reuse, (ii) water conservation through innovations in textile processing machines, (iii) water conservation through innovations in textile processing methods, (iv) water conservation through innovations in textile chemicals and auxiliaries, and (v) tools for processing water use analysis and conservation. However, only several studies on BAT were mentioned briefly while there is no information on the optimization of production scheduling of wet processes.

In this study, the methods of effective planning of the production of the textile industry have been compiled from the studies. There has not been a review on this scope in the literature so far. This review aims to introduce and compare the different perspectives on sustainable production in textile industry via BAT and nonconventional optimization methods, which suggest affordable solutions to reduce the pollution without increasing operating costs. It is also possible to reduce the energy used in some of the applications.

2. Approaches to sustainable production

Sustainable production is important in order to efficient use of resources, reduction of waste, and related costs. There are several reference documents that are suggesting techniques to analyze and modify the textile processes to decrease the consumption of water and energy resulting in the reduction of the pollution.

As a valuable contribution to the literature on the cleaner processes, 18 emerging technologies using the energy and water effectively in the textile industry were explained with their backgrounds, benefits, and commercialization [25]. The technologies were compared with the ability of water saving, energy saving, material saving, and time saving in addition to the reduction in wastewater. As a result, technologies fulfilled all the abilities were enzymatic treatments, ultrasonic treatments, advanced cotton fiber pre-treatment to increase dye receptivity, plasma technology, and foam technology in the finishing process.

In the literature, BAT and metaheuristic optimization methods are studied to provide sustainability and clean production to the textile processes.

2.1 Applications of the best available techniques (BAT)

Alkaya and Demirer studied environmental performance evaluation and sustainable production applications in a woven fabric manufacturing mill in Turkey [26]. The aim of the study was to decrease water consumption, wastewater generation, energy consumption and resulting greenhouse gas emissions, and sodium salt consumption. Baseline data were collected for 8 months. The amount of water consumed per kilogram of product was found as 138.9 L. Additional 4 months spent on implementation and 12 months for monitoring the sustainable production applications. Environmental benchmarking was done by collecting specific resource consumption and waste generation data, which are known as Environmental Performance Indicators (EPIs). As a result of investigation of the process by using BAT, five applications were applied on the process: use of drop-fill washing instead of overflow, reuse of stender cooling water, reuse of singeing cooling water, renovation of water softening system, and renovation of valves and fittings. As a result, amount of wastewater of the process was reduced by 43.4% and CO_2 emission, which is mainly originated from energy consumption, was decreased by 20.2%. Total water consumption was also reduced by 40.2%. Additionally, total energy consumption was decreased by 17.1%.

Ozturk and coworkers investigated a cotton and polyester knitting-weaving fabric and subsequent finishing-dyeing mill in Turkey in terms of BAT applications [5]. The mill with two main production lines had bleaching and dyeing capacities of 2412 and 6682 tons/year, respectively. Freshwater consumed each day was almost 3100 tons before the modification. After the data for 3 years were analyzed, 14 BAT including good management practices, water minimization, and chemical minimization/substitution were chosen from 92 suitable improvements considering mainly their priority, techno-availability, and potential benefits. Some of them are reuse/recovery of washing/rinsing and softening wastewater, reuse of suitable dye bath, caustic recovery from mercerization process wastewaters using membrane process, chemical substitution, and so on. The mill was consuming 95–102 L water per kg product, 9–10 g dyestuff/kg product, and 347–383 g/kg product. After the implementation of selected BATs, the probable reduction in the consumption of water and chemicals was estimated as 43–51 and 16–39%, respectively. The wastewater flow rate was expected to be reduced by 45–52%. Payback period of the implementation was calculated as 26 months at most.

Kocabas and coworkers [6] carried out BAT on a denim manufacturing mill in Turkey. First, they gathered information about the sources of wastewater and their type, quantity and composition as in the BAT reference document (ref BREF).

The consumption of water and energy was recorded from the beginning till the end of the study. Seven operations were evaluated as installing the flowmeters, semi-countercurrent rinsing, reusing the wastewater after treatment, recovering and reusing the wash water from mercerization, reusing the concentrate stream of reverse osmosis plant in sanitary, minimizing the water consumption during regeneration of ion exchangers used in water softening, and reusing the cooling waters in the production process. After implementation of the BAT actions, total water and energy consumption per kg fabric were reduced by 29.5 and 9%, respectively. As a result, saving from the expenses was roughly estimated as EUR 987,000 annually, while total amount of investment was EUR 45,000. There is no information given about the reduction of the amount of wastewater as a result of improvement.

Yukseler and coworkers [27] tried to implement cleaner production to denim manufacturing textile mill in Turkey. The BAT methodology was followed to reduce the amount of consumed water and wastewater generation through the characterization of the wastewater and selection of the wastewater streams to reuse in the process. Selected BAT actions were reusing of wastewaters in the dyeing process after the treatment, recovering of caustic from alkaline finishing wastewaters, reusing of biologically treated composite mill effluent after membrane process, minimizing the wash water consumption in the water softening plant, reusing the concentrate stream from reverse osmosis plant, and reducing water consumption by countercurrent washing in dyeing and finishing processes. As a result, the reduction in the total specific water consumption was evaluated as 30%.

Ozturk and coworkers studied on the improvement of a wool and acrylic fiber production mill in Turkey to a cleaner production by using BAT measures [28]. Suggested BAT actions were reuse of wastewaters from wool yarn softening, LP-VP printing machines and acrylic yarn washing, machinery modifications, reuse of steam condensate, and good management practices. Additionally, replacement of 12 toxic chemicals with biodegradable ones and installing an automatic dosage system were suggested in order to reduce the chemical load in the wastewaters. Energy saving precautions were determined as the implementation of energy recovery systems for high-temperature wastewater flows and flue gas streams, process monitoring control, and various machinery optimizations. After the mass balance calculations, it was estimated that all of the implementations could reduce total water consumption by 35–65%, total energy consumption by 70%, chemical load by 31%, and waste generation by 5–10%. The payback period of the installations was estimated as 4 years.

Kalliala and Talvenmaa investigated the major six textile manufacturers in Finland considering environmental effect of wet processing and suggested appropriate actions of BAT [29]. The study was found especially important since all the industrial manufacturers discharge their wastewater to municipal sewage treatment plants under a strict control in Finland. Energy, water, and chemical consumption data of the processes were collected from the process statistics of the companies. Energy consumption was evaluated between 55 and 152 MJ/kg product, while water consumption was between 144 and 380 L/kg product, and the CO_2 emission was found between 3484 and 8937 g/kg product. A detailed chemical consumption table was also prepared for the study. As a result, suggested BAT actions were planned as the application of automatic dosing of chemicals and dyes, effective use of equipment capacity, recycling and monitoring of process water and energy used, recovery and purification of process liquor, monitoring of wastewater with toxicity analyses, and monitoring of flue gas emissions.

A LIFE funded project BATTLE (05 ENV/IT/000846) was proposed to evaluate the applicability of BAT such as those described in the textile reference documents (BREFs) for the implementation of the European Directive IPPC 96/91/CE to small-medium enterprises (SMEs) in terms of technical and economical feasibility of water recycling for European textile sector [30]. In the project, a prototype application was included for cleaner production in a textile dyeing and printing company. After the environmental impact and process data were evaluated, the process was modified to recover and reuse the process water. As a result, the amount of total recovered water was determined as 7691 m^3/month with an appropriate quality. The reduced values in the water consumption, energy consumption, wastewater, and greenhouse gas generation given in the literature are tabulated in **Table 1**.

2.2 Metaheuristics

Using nonconventional optimization methods in the textile industry by considering both the delivery date and environmental issues is a quiet new area in the literature. Although the major part of these studies is accumulated on the scheduling of the dyeing process, novel optimization and decision-making algorithms (AI—Artificial Intelligence) have a huge potential in a large area of textile production such as cotton grading, yarn CPS (count strength product) prediction, yarn grading, fabric colorfastness grading, fabric comfort, and fabric inspection [31].

The studies on the nonconventional optimization methods can be classified into two groups: studies on the development of hybrid metaheuristic algorithms and the studies on the application of the genetic algorithm in real processes.

2.2.1 Developed hybrid algorithms

Huynh and Chien [4] worked on the parallel batch processing machine scheduling with sequence-dependent setup, arbitrary job size, different due date, and incompatible job family. They proposed a multi-subpopulation genetic algorithm with heuristics embedded (MSGA-H) to improve batching and scheduling simultaneously. The validity of the algorithm was tested by an empirical study with data supplied from a textile dyeing manufacturing in Taiwan. The results have shown the practical viability of the proposed MSGA-H. The reduction of used water and wastewater were not reported in this study.

In another study on the textile dyeing process [20], parallel machine scheduling problem with environmental requirements and tardiness were solved by generating a multi-objective genetic algorithm with tabu-enhanced local search (MOGA-TIG). Three objective functions were defined to obtain a sustainable schedule: the number

	Reduction in				
Mill/Process	**Wastewater (%)**	**Total water consumption (%)**	**Flue gas (%)**	**Total energy consumption (%)**	**Ref.**
Cotton and polyester fabric/ Finishing and dyeing	45–52	49–55v	—	—	[5]
Denim/Dyeing and finishing processes	—	29.5	—	9	[6]
Woven fabric/Dyeing, finishing, cooling, and utilities	43.4	40.2	20.2	17.1	[26]
Denim/Waste-water treatment	—	30	—	—	[27]
Wool and acrylic fiber/Dyeing and finishing	—	35–65	25–65	70	[28]

Table 1.
The reduction values in the water consumption, energy consumption, wastewater, and greenhouse gas generation.

of setups to be minimized to reduce the water consumption and use of detergent caused by the changing of the job, the utilization of the machines to be maximized to prevent the pollution caused by ineffective usage of the batch capacity, and tardiness that may cause penalty on late completion of the jobs to be minimized. When the performance of the algorithm was compared with the latest multi-objective evolutionary algorithms in the literature, it was concluded that the suggested algorithm gives a satisfactory solution in a short time. However, since the algorithm was not applied on a real industrial process, there was no data available for the reduction of the used water and dye in this study.

Zhang and coworkers studied on a similar process with the previous study, and they approached to the problem as bi-objective optimization model [21]. The objectives of the study were to reduce delivery tardiness cost and pollutant emission caused by the cleaning operation of setups before each job. Multi-objective particle swarm optimization algorithm enhanced by problem-specific local search

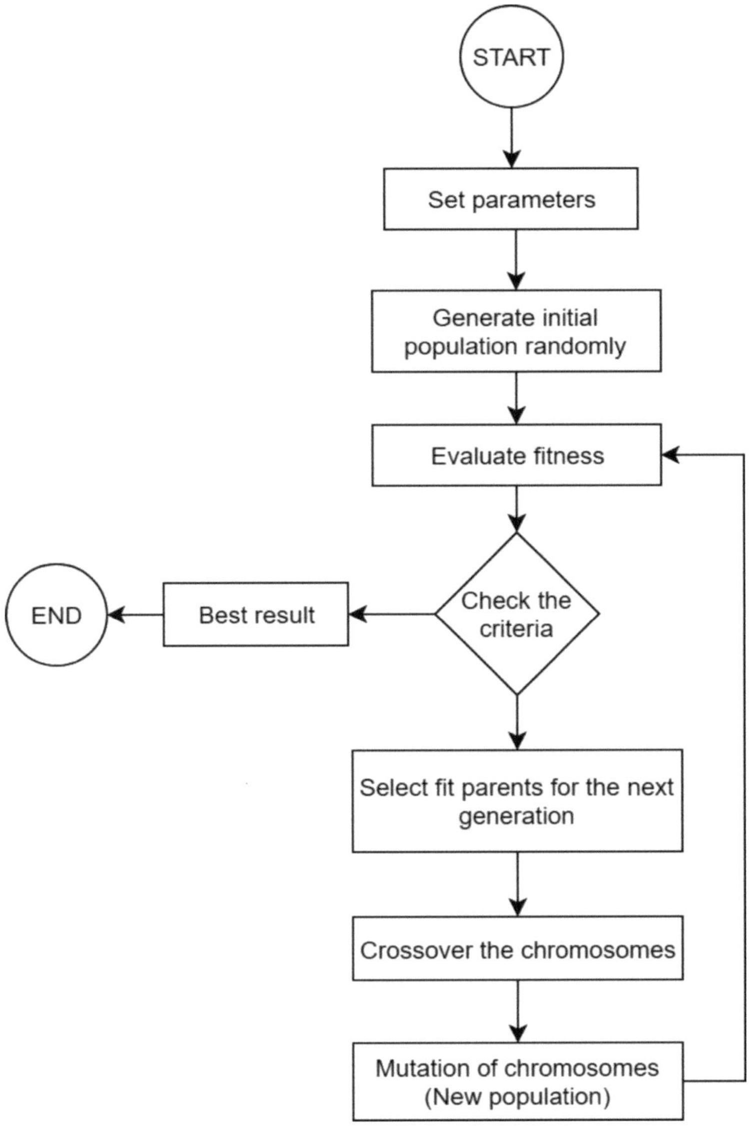

Figure 1.
Flowchart of the genetic algorithm [32].

(MO-PSO-L) technique was utilized to solve the problem. The algorithm was verified on computational experiments by using simplified realistic production data and also compared with the universal multi-objective optimizer NSGA-II and multi-objective imperialist competitive algorithm (MOICA). The performance of the algorithm was measured by using three indicators: the overall nondominated vector generation (ONVG), the C-metric (CM), and the Tan's spacing (TS). It was possible to conclude that MO-PSO-L algorithm found more Pareto solutions and became superior to other algorithms in terms of uniform distribution of the solutions.

2.2.2 Application of the genetic algorithms

Zhou and coworkers [22] used the genetic algorithm to schedule a dyeing process. Scheduling objectives are delivery time of the products, complete filling of the dyeing vessel, putting the same type, color, depth and production process of cloth in the same dyeing vessel, and sorting the depth of color such as from light to dark. Genetic algorithm was started with assigning all orders to different dye vats and producing a number of individuals. Each individual was checked to meet the delivery date. Then, fitness function was computed for the population. After the crossover and mutation steps, next generation was chosen according to the parent and child fitness. The procedure repeated itself until reaching the maximum number of evolutionary. The algorithm was verified on the data supplied from an enterprise in China. It was reported that the optimization of the schedule reduced the freshwater consumption about 18.4–21%, also resulting less wastewater.

The flowchart of a typical genetic algorithm is shown in **Figure 1**.

Jiang and coworkers applied genetic algorithm to the data obtained from a real process in China to optimize production schedule [23]. The aim of the study was to reduce the production time and the volume of freshwater consumed in the dyeing operation. The results showed that the optimized schedule could reduce the production time as 10–15% while reducing the volumes of freshwater and wastewater as 20–30 and 20%, respectively. Additionally, it was found that adding alternative production lines (from 1 to 4) in the process shortens the total production time and lessens the amount of wastewater.

3. Conclusion

Textile industry, which is dominated by small and medium enterprises (SMEs), has a wide variety of products with different colors, fabrics, and fabric types by its nature. Additionally, wet processes consume large amount of water, energy, and chemicals, which are expensive to separate in the treatment processes. In the last decades, the studies with environmental and financial considerations in the textile industry have been increased, and they offered cleaner production approach, such as best available techniques (BAT) referred by the European Council and the nonconventional (metaheuristic) optimization methods, instead of end-of-pipe approach.

From the literature, it was revealed that BAT actions offer a substantial water and energy savings up to 65 and 70%, respectively. However, many of the studies involve potential implementation results, not the real ones. Therefore, more implementation studies should be conducted on actual processes in order to encourage the other enterprises.

Through the age of Industry 4.0, the enterprises, which use intelligent techniques and effective planning, will survive. Consequently, the studies on the optimization of the dyeing process scheduling have become more complex and multi-objective to be solved with nonconventional optimization methods. In the literature, only a few studies were found because of the novelty of the subject. However, those studies were indicated a remarkable environmental benefit while reducing the costs of the wastewater treatment process. The reduction of water consumption was obtained up to 30% without any installations to the process.

When the literature on BAT and metaheuristic studies are compared, it is seen that the studies used BAT has more industrial sense and feedback; however, just insufficient number of studies that were used metaheuristic optimization quantified the improvement on the studied processes. The information on the application of the novel optimization methods on the actual industrial applications is necessary. Therefore, future studies on this subject will be fruitful for the literature.

Author details

Miray Emreol Gönlügür
Chemical Engineering Department, Ege University, Izmir, Turkey

*Address all correspondence to: mirayemreol@gmail.com

References

[1] Gümüş D, Akbal F. Photocatalytic degradation of textile dye and wastewater. Water, Air, & Soil Pollution. 2011;**216**:117-124. DOI: 10.1007/s11270-010-0520-z

[2] Takahashi N, Kumagai T. Removal of dissolved organic carbon and color from dyeing wastewater. Ozone Science and Engineering. 2006;**28**(4):199-205. DOI: 10.1080/01919510600704056

[3] Chen L, Wang L, Wu X, Ding X. A process-level water conservation and pollution control performance evaluation tool of cleaner production technology in textile industry. Journal of Cleaner Production. 2017;**143**:1137-1143. DOI: 10.1016/j.jclepro.2016.12.006

[4] Huynh NT, Chien CF. A hybrid multi-subpopulation genetic algorithm for textile batch dyeing scheduling and an empirical study. Computers and Industrial Engineering. 2018;**125**:615-627. DOI: 10.1016/j.cie.2018.01.005

[5] Ozturk E, Koseoglu H, Karaboyaci M, Yigit NO, Yetis U, Kitis M. Minimization of water and chemical use in a cotton/polyester fabric dyeing textile mill. Journal of Cleaner Production. 2016;**130**:92-102. DOI: 10.1016/j.jclepro.2016.01.080

[6] Kocabas AM, Yukseler H, Dilek FB, Yetis U. Adoption of European Union's IPPC Directive to a textile mill: Analysis of water and energy consumption. Journal of Environmental Management. 2009;**91**(1):102-113. DOI: /10.1016/j.jenvman.2009.07.012

[7] Brik M, Schoeberl P, Chamam B, Braun R, Fuchs W. Advanced treatment of textile wastewater towards reuse using a membrane bioreactor. Process Biochemistry. 2006;**41**(8):1751-1757

[8] Nishitani K, Kaneko S, Fujii H, Komatsu S. Effects of the reduction of pollution emissions on the economic performance of firms: An empirical analysis focusing on demand and productivity. Journal of Cleaner Production. 2011;**19**(17-18):1956-1964. DOI: 10.1016/j.jclepro.2011.06.021

[9] United Nations Environment Programme (UNEP). Green Economy and Trade, United Nations Environment Programme Trends, Challenges and Opportunities; 2013

[10] Alkaya E, Böğürcü M, Ulutaş F, Demirer GN. Cleaner (sustainable) production in textile wet processing. Non-Conventional Textile Waste Water Treatment. 2012. pp. 205-229. Available from: https://www.scopus.com/inward/record.uri?eid=2-s2.0-84895285170&partnerID=40&md5=01805914979924a3f86da9e57ba84638

[11] European Integrated Pollution Prevention and Control Bureau (EIPPCB). Reference Document on Best Available Techniques for the Textiles Industry. 2003. Available from: http://eippcb.jrc.es/reference/BREF/IS_Adopted_03_2012.pdf http://eippcb.jrc.ec.europa.eu/reference/ [Accessed: 28-12-2018]

[12] Camargo VCB. Optimization of processes in textile industry: Model and solution methods [thesis]. University of Porto; 2012

[13] Çelekli A, Bozkurt H, Geyik F. Use of artificial neural networks and genetic algorithms for prediction of sorption of an azo-metal complex dye onto lentil straw. Bioresource Technology. 2013;**129**:396-401

[14] Schubert M, Muffler A, Mourad S. The use of a radial basis neural network and genetic algorithm for improving the efficiency of laccase-mediated dye decolourization. Journal of Biotechnology. 2012;**161**(4):429-436. DOI: 10.1016/j.jbiotec.2012.08.003

[15] Al-Obaidi MA, Li JP, Kara-Zaïtri C, Mujtaba IM. Optimisation of reverse osmosis based wastewater treatment system for the removal of chlorophenol using genetic algorithms. Chemical Engineering Journal. 2017;**316**:91-100

[16] Bagheri M, Mirbagheri SA, Bagheri Z, Kamarkhani AM. Modeling and optimization of activated sludge bulking for a real wastewater treatment plant using hybrid artificial neural networks-genetic algorithm approach. Process Safety and Environment Protection. 2015;**95**:12-25. DOI: 10.1016/j.psep.2015.02.008

[17] O'Brien M, Mack J, Lennox B, Lovett D, Wall A. Model predictive control of an activated sludge process: A case study. Control Engineering Practice. 2011;**19**(1):54-61. DOI: 10.1016/j.conengprac.2010.09.001

[18] Ghaedi AM, Vafaei A. Applications of artificial neural networks for adsorption removal of dyes from aqueous solution: A review. Advances in Colloid and Interface Science. 2017;**245**:20-39. DOI: 10.1016/j.cis.2017.04.015

[19] Nasab SG, Semnani A, Teimouri A, Yazd MJ, Isfahani TM, Habibollahi S. Decolorization of crystal violet from aqueous solutions by a novel adsorbent chitosan/nanodiopside using response surface methodology and artificial neural network-genetic algorithm. International Journal of Biological Macromolecules. 2018;**124**:429-443. DOI: 10.1016/j.ijbiomac.2018.11.148

[20] Zhang R. Sustainable scheduling of cloth production processes by multi-objective genetic algorithm with tabu-enhanced local search. Sustainability. 2017;**9**(10):1754

[21] Zhang R, Chang PC, Song S, Wu C. Local search enhanced multi-objective PSO algorithm for scheduling textile production processes with environmental considerations. Applied Soft Computing. 2017;**61**:447-467. DOI: 10.1016/j.asoc.2017.08.013

[22] Zhou L, Xu K, Cheng X, Xu Y, Jia Q. Study on optimizing production scheduling for water-saving in textile dyeing industry. Journal of Cleaner Production. 2017;**141**:721-727. DOI: 10.1016/j.jclepro.2016.09.047

[23] Jiang W, Yuan Z, Bi J, Sun L. Conserving water by optimizing production schedules in the dyeing industry. Journal of Cleaner Production. 2010;**18**(16-17):1696-1702. DOI: 10.1016/j.jclepro.2010.07.004

[24] Hussain T, Wahab A. A critical review of the current water conservation practices in textile wet processing. Journal of Cleaner Production. 2018;**198**:806-819. DOI: 10.1016/j.jclepro.2018.07.051

[25] Hasanbeigi A, Price L. A technical review of emerging technologies for energy and water efficiency and pollution reduction in the textile industry. Journal of Cleaner Production. 2015;**95**:30-44. DOI: 10.1016/j.jclepro.2015.02.079

[26] Alkaya E, Demirer GN. Sustainable textile production: A case study from a woven fabric manufacturing mill in Turkey. Journal of Cleaner Production. 2014;**65**:595-603. DOI: 10.1016/j.jclepro.2013.07.008

[27] Yukseler H, Uzal N, Sahinkaya E, Kitis M, Dilek FB, Yetis U. Analysis of the best available techniques for wastewaters from a denim manufacturing textile mill. Journal of Environmental Management. 2017;**203**:1118-1125. DOI: 10.1016/j.jenvman.2017.03.041

[28] Ozturk E, Karaboyaci M, Yetis U, Yigit NO, Kitis M. Evaluation of integrated pollution prevention control in a textile fiber production and dyeing

mill. Journal of Cleaner Production. 2015;**88**:116-124

[29] Kalliala E, Talvenmaa P. Environmental profile of textile wet processing in Finland. Journal of Cleaner Production. 2000;**8**(2):143-154

[30] BATTLE Project Final Report (LIFE05 ENV/IT/000846) [Internet]. 2009. Available from: http://www.life-battle.bologna.enea.it/files/FINAL Report.pdf [Accessed: 08-01-2019]

[31] Doke SS, Shanmugam N. Artificial intelligence and its applications in textile. Asian Textile Journal. 2002;**11**(7):49-54

[32] Heidari N. Fabric textile defect detection, by selecting a suitable subset of wavelet coefficients, through genetic algorithm. International Journal of Image Processing. 2011;**5**(1):25-35

Chapter 2

Use of Ozone in the Textile Industry

Ayşegül Körlü

Abstract

Wet processing of textile materials consumes a large amount of electricity, fuel, and water. Therefore, greenhouse gas emissions and contaminated effluent are environmental problem. The most of the governments in the world warn all the industrial sectors containing textile manufacturing to be careful about environmental pollution. Increasing in public awareness of environment and competitive global market forces the textile industry to manufacture textile products environmentally. Environmental pollution in textile wet processes can be reduced by four main ways. They are process optimization (reducing in water, chemical energy consumption, and time loss), use of ecofriendly chemicals, reuse of water, and new technologies like ozone and plasma technologies, transfer printing, enzymatic processes, etc. This chapter is about the use of ozone in the textile industry.

Keywords: textile, wet processing, ozone, ozone treatment, environment

1. Introduction

The aim of wet processes is to improve the appearance, texture, or performance of a textile material. Wet processes consist of pretreatment, dyeing, printing, and finishing processes. Wet processes compose of application of chemicals, fixation, washing, and drying stages generally [1, 2].

The textile dyeing and finishing industry is one of the largest water users. It uses huge amount of water throughout all processing operations [2]. A lot of dyestuffs, chemicals, auxiliaries, etc. are applied to textile materials in water baths. Ecological problems are commonly related with water contamination in the textile dyeing and finishing industry. Wastewater of the textile industry is hot, has strong odor, and colored by dyestuffs, which are used as a part of dyeing and/or printing process [2–4].

Clean production methods consider all possibilities that will lessen the effects of pollution problem in wet processing of textiles and will save water and/or energy. The main methods are below [5]:

1. Process optimization (water and time saving in every possible area, use of less chemical, working in lower temperatures).

2. Use of environmentally friendly chemicals.

IntechOpen

3. Reuse of water.

4. New technologies in wet processing (enzymatic processes, ultrasound, ultraviolet, plasma and ozone technologies, dyeing in CO_2-containing environment, etc.).

In this chapter, the use of ozone as some ecofriendly production method in wet processing of textiles is investigated.

2. Ozone

Ozone is a strong oxidant agent, which can be produced synthetically, as well as is being naturally available in the atmosphere. Ozone layer behaves like a shield against ultraviolet radiation. Because it absorbs UVB and UVC light during the cycle (**Figure 1**) of formation and destruction of ozone in the atmosphere [6–8].

Christian Schönbein described "ozein" odor during electrolysis of water in 1839. Thomas Andrews found out that ozone was formed only by oxygen in 1856. In 1863, Soret defined the relationship between oxygen and ozone. He determined that 3 volumes of oxygen produce 2 volumes of ozone. Ozone is thermodynamically unstable and spontaneously reverts to oxygen (**Figure 2**). It dissolves very quickly in pure water and respects for Henry's law. Ozone immediately reacts with inorganic and organic substances dissolved in biological water generating a variety of free radicals [9, 10].

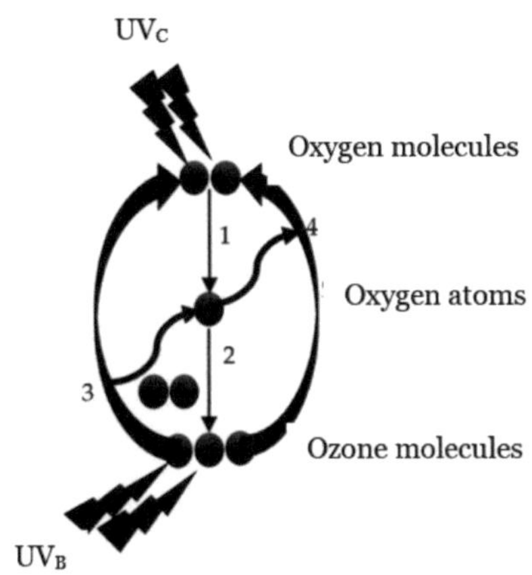

Formation : $O_2 + UV_C \rightarrow O + O$

$O + O_2 \rightarrow O_3$

Destruction: $O_3 + UV_B \rightarrow O_2 + O$

$O + O_3 \rightarrow 2O_2$

Figure 1.
Cycle of formation and destruction of ozone [8].

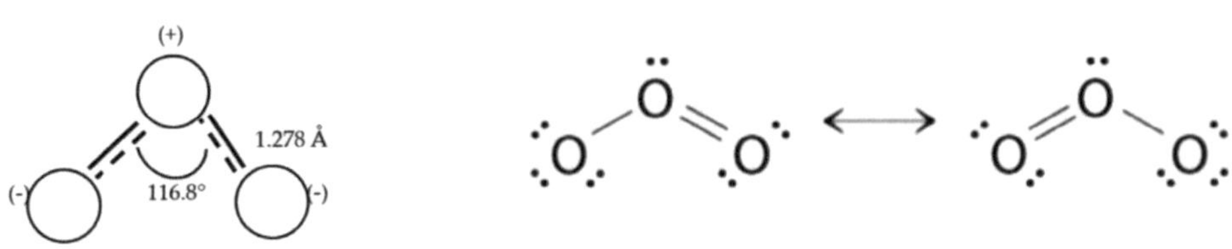

Figure 2.
Ozone molecule [6, 11]. $3O_2 \rightleftharpoons 2O_3$ $\Delta H_f{}^\circ$ at 1atm =+284,5 kJ.mol^{-1}

2.1 Generation of ozone

Ozone must be generated "in situ" because it is very reactive gas and cannot be stored and transported to anywhere. So, it has to be generated [3]. The basic methods for generating ozone artificially are below [5–7, 9, 12–15]:

- **Photochemical ozone generation:** Oxygen atoms formed by the photodissociation of oxygen by short-wavelength UV radiation react with oxygen molecules to form ozone. The theoretical quantum yield of ozone by photochemical technique is 2%. Nevertheless, the actual yield is approximately 0.5% in practice. Because, the low-pressure mercury lamps produce not only the 185-nm radiation responsible for the production of ozone, but also the 254-nm radiations that destroy ozone. Medium-pressure UV produces higher levels of 185-nm radiation, and it generates more ozone. The low concentrations of ozone from UV generators limit their usage for water treatment to special applications. But it can be used in air treatment effectively.

- **Electrolytic ozone generation:** An electrolytic cell is used for electrolytic ozone generation. Electrolysis involves converting oxygen in the water to ozone by passing the water through positively and negatively charged surfaces. Electrolysis of water can generate high concentrations of ozone. However, the output is low and the method is more expensive than the corona discharge process. Small electrolytic units can be used for treatment of ultrahigh purity waters in pharmaceutical and electronic industries.

- **Radiochemical ozone generation:** High-energy irradiation of gaseous or liquid oxygen by radioactive rays can help the formation of ozone. Energy efficiency of the method is greater than that of ozone produced by electric discharge. However, it has not yet been commercialized due to complex structure, problems associated with recovery of ozone, and separation of by-products and radioactive material.

- **Ozone generation by corona discharge (silent electrical discharge):** Ozone is generated by providing air or oxygen gas into the generator. And oxygen or air is converted into ozone by the electric discharge. Primary components in air are firstly separated into reactive atoms or radicals by effect of the intense electric field. Then, these reactive atoms can react among themselves. Ozone generation by corona discharge is especially the most widely used method for water treatment.

2.2 Measurement methods

It is necessary to determine the concentration of ozone produced by an ozone generator because of efficiency of processes, costs, excessive ozone, and environmental drawback [14, 16]. Many analytical methods for the determination of ozone concentration have been described in the literature. However, most of them are not specific and often give incorrect results [17]. Analysis of ozone is difficult because of the instability of pure ozone, volatilization from solution, the rapid decomposition of ozone in water, and the reaction with trace contaminants in water, etc. [18].

Ozone can be analyzed by methods given below [15]:

- titrimetry,

- direct and colorimetric spectrometry,

- amperometry,
- oxidation-reduction potential (ORP),
- chemiluminescence,
- calorimetry,
- thermal conductivity,
- gas-phase titration with NO, and
- isothermal pressure change on decomposition.

The last four methods are not commonly used for analysis. It is necessary to be aware of its reactivity, instability, volatility, and the potential effect of interfering substances to measure the amount of ozone in water correctly. Ozone sometimes is sprinkled in drops by using an inert gas for analysis in the gas phase or on reabsorption in a clean solution in order to eliminate interferences [14].

2.2.1 Iodometric methods

Gaseous ozone from ozone generator is absorbed by aqueous potassium iodide solution. On the other hand, a preformed ozone solution can be alternatively treated with aqueous potassium iodide solution. The liberated iodine is measured by spectrometer or titration with sodium thiosulfate. The pH value of iodine solution is adjusted to 2. Then, it is titrated with titrant solution sodium thiosulfate and starch indicator [14, 19–21]. Theoretically, one molecule of ozone releases one molecule of iodine as the triiodide ion. It is a standard method [21]. Oxidants like H_2O_2 and NO_x are problem for the measurement method because of their interference with the analysis. The method is sensitive to pH, buffer composition, buffer concentration, iodide concentration, sampling techniques, temperature, and time [14, 19, 21].

Aqueous ozone solution is added to 2% potassium iodide in 0.1 M neutral phosphate buffer containing a known amount of arsenic (III). And then, the excess As (III) is back-titrated with standard iodine using a starch end point [21, 22].

Standard method for residual chlorine analysis is adapted for residual ozone as Palin DPD (N,N-diethyl-p-phenylenediamine) method. Ozone oxidizes iodine ion in phosphate buffer of pH 6.4. Then, the released iodine oxidizes DPD, and it is measured colorimetrically or is titrated with standard ferrous ammonium sulfate due to the formation of pink Wurster cation [21].

The other adapted standard way is amperometric method. Ozone oxidizes iodide ion in acetate buffer of pH 4.0–4.5 in the presence of sodium thiosulfate, phenylarsine oxide (PAO), or inorganic arsenic (III). These reagents are titrated with standard iodine to an amperometric end point without acidification. Grundwell et al. compared (**Table 1**) the four currently popular iodometric methods for aqueous ozone analysis in 2008 [21].

2.2.2 Direct spectrometry

Ozone has a peak absorption at 254 nm wavelength, and it is the ultraviolet spectrum, so absorbance of the gaseous ozone can be measured at 254 nm by direct UV spectrometry. The method is sensitive in the molar absorptivity. Interference of CO, hydrocarbons, NO_x, or H_2O vapor has no noticeable effects on measurement.

Method	O_3 reduction		I_2 titration
	Excess reagent	**pH**	**pH**
Iodometric	None	3.5–9	2
Amperometric	S_2O_3 = PAO or As (III)	4.5	4.2–6.8
As (III) back	As (III)	6.8	6.8
DPD	DPD	6.4	6.4

Table 1.
Iodometric methods for ozone analysis [21].

UV absorbance measurement method is mainly useful for gas analysis. This method also can be used for measuring aqueous ozone solution. However, interference from turbidity, dissolved inorganics, and organics is a problem. Ozone is sprinkled into the gas phase in order to eliminate interference for measurement. If the sample is liquid, the best sample for this method will be clean water free of UV-absorbing impurities [14, 19, 23–25].

2.2.3 Colorimetric spectrometry

Several different colorimetric methods are used for measuring ozone residuals. But most of them are sensitive to significant interference from secondary oxidants [14].

The first reagents for measuring ozone in air and exhaust gases are indigo and its water-soluble derivatives, the sulfonated indigo compounds like indigo disulfonate, indigo carmine, and indigo trisulfonate. Water-soluble derivatives of indigo, indigo disulfonate, and indigo trisulfonate are pH or redox indicators. Purity and age of the indigo trisulfonate are very important for the method because they affect the stoichiometry of the reaction [17, 19].

The indigo molecule contains only one double bond (C=C). It reacts with ozone in order to produce sulfonated isatin and similar substance (**Figure 3**). Maximum absorbance of indigo is at 600 nm [17, 26]. If the pH value is below 4, sensitivity of the method does not vary with ozone concentration, small changes of temperature of reaction, or the chemical composition of the water. The advantage of the method is applicable for lake water, too hard groundwaters, and biologically treated domestic wastewater. Indigo trisulfonate method is quantitative, selective, fast, and simple. Classical instrumentation of water work laboratories is enough for measurement. The method is based on the decolorization of the dye by ozone. The loss of color is directly proportional to the concentration of ozone, and pH value of the sample is adjusted to about 2 in order to minimize destruction of the ozone by hydroxide ions [17, 19, 21]. The concentration is the difference in absorbance between sample and blank [16]. Mn^{+2} ion is a problem in this method. Because, oxidation products from the reaction of Mn^{+2} ions with ozone can demolish indigo trisulfonate. So, glycine is added to the sample in order to demolish the ozone

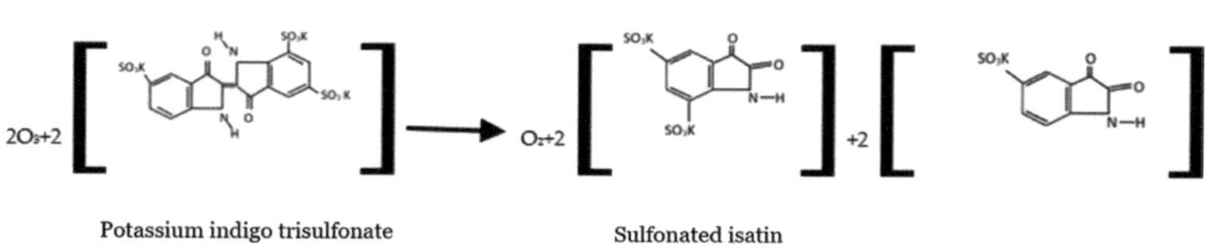

Figure 3.
Ozonation of potassium indigo trisulfonate [26].

selectively. Then, indigo reagent is added to the sample to measure the seeming ozone concentration because of the reaction with manganous ion oxidation products. This value is subtracted from the value of the sample without glycine [19].

Indigo method is applied to "AccuVac Ampul" ozone measurement. For the measurement, all the chemicals are packaged under vacuum into an ampule. The ampule is like a cuvette, and it is used precisely for the spectrophotometric determination of the dissolved ozone [25].

2.2.4 Electrochemical methods

These methods can include amperometric titration, which is mentioned under the title of iodometric methods. Amperometric analyzers use an electrochemical cell to determine the ozone. There are two types of analyzer [25]:

- The bare-metal electrodes, which are in direct contact with the water;
- The electrochemical cell separated from the process water by a semipermeable membrane.

In the literature, there are various electrochemical methods using a solid redox polymer electrolyte-based amperometric sensor, a lignin-modified, glassy carbon electrode, and multiwalled carbon nanotubes to analyze ozone [16].

General advantages of electrochemical methods are low cost, easy operation, potential for miniaturizing and automation, simple portable devices for fast screening purposes, and in-field/onsite monitoring [16]. But every 1 or 2 months, amperometric analyzers have to be calibrated against a reference method such as UV, or the colorimetric indigo methods [25].

2.2.5 Chemiluminescent methods

Light is produced on the reaction of ozone with ethylene in gas phase. It is chemiluminescence. This type of analyzer is based on chemiluminescence and it is measured with a photomultiplier tube. It is comparable to the ozone concentration. Aqueous solutions of ozone emit light on the reaction of ozone with miscellaneous dyes. They are benzoflavin, acridine yellow, indigo trisulfate, fluorescein, etc. [14].

3. Ozone applications in the textile industry

There are two types of ozone application in the textile industry. They are:

- application in the gas phase and
- application in aqueous phase.

For wet processes, aqueous ozone is more practical than gaseous ozone because working principle of finishing machines is suitable for solution. Use of gaseous ozone needs special airproof machines because of occupational health and comfort. Material of leakproof gasket has to resist gaseous ozone.

Gas phase: Half-life of gaseous ozone is more than that of aqueous ozone. So, it decomposes slowly [13]. Gaseous ozone is easily affected by the catalyst. Light, trace organic matter, nitrogen oxides, mercury vapor, and peroxides act as catalysts for homogeneous catalysis. Metals and metal oxides are catalysts for heterogeneous catalysis. If there is no any effect of the catalyst on the mixture, it will be stable. Porous solid substrates can adsorb gaseous ozone [14]. Textiles are the sample of porous solid substrates.

Aqueous phase: Solubility of ozone in water is better than the oxygen [27]. However, ozone dissolves moderately in water. It follows Henry's law [13]. The solubility of ozone depends on pressure, temperature, and ionic strength [14]. Efficient moving of ozone to solution needs dispersion of gaseous ozone into small bubbles. Positive-pressure ozone contractors, negative-pressure reactors (Venturi), and injectors achieve it [13]. At room temperature, decomposition of ozone in pure water is very slow. However, ozone decomposes with the catalysts like hydroxyl ion, trace metals, H_2O_2/HO_2^-, organic substances, heat, and UV light [14].

The parameters affecting the physical mass-transfer rate of ozone into water are:

- gaseous ozone concentration,
- temperature,
- pressure,
- solution composition (pH, ionic strength, reactive substances),
- gas dispersion,
- turbulence,
- type of contactor [14].

Recently, scientific studies about the use of ozone in textile manufacturing are very popular. But the use of ozone in textile manufacturing is not common in practice. It is only commercially common in denim and garment washing. Therefore, the use of ozone in denim washing is discussed in a special title. The other titles are based on fiber types.

3.1 Use of ozone in denim washing

The denim garments are very popular, and they are preferred by people of all ages, classes, and genders. Depending on the desired effect, denim garments are treated with different substances [28, 29]. A lot of dry and wet processing techniques are used for desired effects [30]. Wet processes in denim washing are not environment friendly. High water and energy consumption, large amount of wastewater, and solid waste like pumice stones are generally environmental problem in denim washing [29, 30]. Sodium hypochlorite is a very common bleaching agent in denim washing. Especially AOXs (adsorbable halogenated organic compounds) are the most important environmental problem. Therefore, chlorine-free bleaching technologies are a good solution for AOX. Ozonation is an alternative bleaching method [28]. Ozonation is a simple and "green" process because it does not require steam and water. Therefore, it greatly reduces process time, water, chemical and energy consumption, and amount of wastewater [31, 32]. Ozone decomposes indigo and other dyes because of high oxidation potential. In addition to denim washing, it is generally applicable to treatment of other textiles like T-shirts, shirts, chinos, and casual wear. In ozonation, the ozone generated in the equipment can commercially provide bleaching effect. It is like washing machine without water for fading of color [31]. And ozone is generally applied to whole of the garment in this ozonation machine. However, local bleached spots on the fabric can also be created by ozone. Ozone gas is scattered onto the denim fabric at a controlled velocity [32].

Özdemir et al. studied on ozonation parameters of denim fabric. They used prewashed denim fabric for ozonation. They inform that water content of the denim fabric is very important for efficiency of ozonation, and 50–60% water pick up value (W.P.V.) is the best for bleaching efficiency. The higher W.P.V. affects the bleaching efficiency of ozonation negatively. For ozonation, acidic and neutral pH values are better than basic pH value. Temperature is one of the most important parameters because higher temperatures decrease the half-life of ozone. Ozonation of wet denim is especially for bleaching [28].

Hmida and Ladhari studied the parameters affecting dry and wet ozone bleaching of denim fabric. Their results about the effect of W.P.A. on bleaching efficiency are also compatible with Özdemir's results. They claim that water film covers the surface of the fabric and swelling of the fibers is achieved. Then, ozone can penetrate into fibers, and bleaching efficiency on wet denim is better than that of dry denim. On the other hand, backstaining problem is solved by ozonation of dry denim [31, 33].

In the other study, denim fabric was treated with the combination of ozonated water, ultrasound, and hydrogen peroxide. According to the results, ozone is more effective with the aid of ultrasonic energy because the ultrasonic cavitations improve the penetration of ozone into the fabric, and then, ozone decomposes indigo [5, 32].

Bağıran et al. compared ozone to other bleaching agents in denim washing. According to results, ozone is one of the strong agents. It follows potassium permanganate and benzoyl peroxide. But, the most important advantage of ozone is environment friendly, and it is a good alternative to the others. Benzoyl peroxide and ozone are the causes of gray tint in bleached denim fabric. The others are the reason for blue tint of bleached denim. After ozonation, loss of strength is not too high because ozone is unstable, and it decomposes indigo primarily [34].

In practice, ozone is applied to garments. However, there are a few studies on ozonation of yarn [35, 36]. Beşen and Balcı try to fade indigo-dyed yarn before weaving and garment processes. The indigo-dyed yarn is ozonated in hank form. Their results about the effect of the ozonation condition on bleaching efficiency are generally compatible with Özdemir's results. On the other hand, the origin of the raw material directly affects the fading degree of the yarn. According to the results, the count of the yarn is the most important parameter on the decrease in strength of the yarns depending on the ozonation process. However, strength loss is not so important. As a conclusion, the ozonation condition can be determined according to desired effect from the yarns. They claim that different fading effects can be achieved by ozonation before weaving process [35].

He et al. investigated effect of ozone on three typical denim yarns (cotton, lyocell, and polyethylene terephthalate (PET)) during the color-fading process. They claim that ozone only smoothly impacts the crystalline structures of these yarns. PET is not affected by ozone because of its aggregate structure. This structure prevents the oxidation and decomposition of PET. They suggest that ozonation for cotton, lyocell, or other cellulosic yarns should be limited within 10 min at the $pH > 7$ with a careful selection of water content [36].

3.2 Ozonation of cellulosic fibers

A lot of researchers study on ozonation of cellulosic fibers, especially cotton. Perinçek et al. studied on the use of ozone gas in bleaching cotton fabrics. Cotton fabric containing 60% water at pH 7 can be bleached in a short time with ozonation treatment. Room temperature is the optimum for ozonation. After ozonation in a short time, the whiteness of the fabric is acceptable for dyeing and DP losses are not so important [5, 10].

Eren and Öztürk also investigated ozonation of cotton fabrics. According to their results, the starch size removal of the greige cotton samples and the water absorbency of the greige and desized cotton samples are increased by ozonation. But ozonation does not remove the motes successfully. Bleaching effect of ozonation is successful because of high oxidation potential of ozone [37].

Maqsood et al. suggest ozonation of cotton fiber for medical textiles and production of nanocrystalline cellulose or nanofibrils of cellulose in their paper [38].

Turhan and Soydaş discussed ozonation of cotton terry fabrics. As the results of the study, ozone cannot sufficiently remove impurities like sizing agents, natural waxes, and oils. Therefore, they suggest desizing the terry fabric before ozonation and rinsing the fabric after ozonation [39].

Perinçek et al. also investigated the effects of new advanced processes on cotton woven fabric. The new advanced processes contain ozonation, ultrasound, and ultraviolet. In this study, cotton fabrics are bleached by combining ozone with ultrasound and ultraviolet. According to results, advanced processes can be used in pretreatment of cotton fabrics. However, advanced processes at 63–65°C are not sufficient for desizing the cotton fabrics and need desizing agents. After ultrasonic treatment followed by ozonation, whiteness and hydrophilicity of the fabric are sufficient for dyeing. The combination of ozone and ultraviolet processes for high whiteness of the fabric is recommended. However, breaking strengths and hydrophilicity values have to be considered carefully [3].

Perinçek recommended a removing method of optical brightener from cotton fabric in her paper. It is difficult to remove optical brightener efficiently from the fabric when any problem is seen on the fabric due to high stability of optical brightener. Hazardous chemicals are generally used to remove it. Therefore, ozonation process has ecological advantage. The results show that ozonation can be used for decolorizing the optical bleached samples. Increasing ozonation time increases the efficiency. However, bursting strength loss of fabric should be taken into consideration due to the oxidation of cellulose. Meanwhile, a new patterning method for optical bleached fabrics is proposed in the paper. It provides fashionable products (**Figure 4**) like batik or tie-dyed cloths [40].

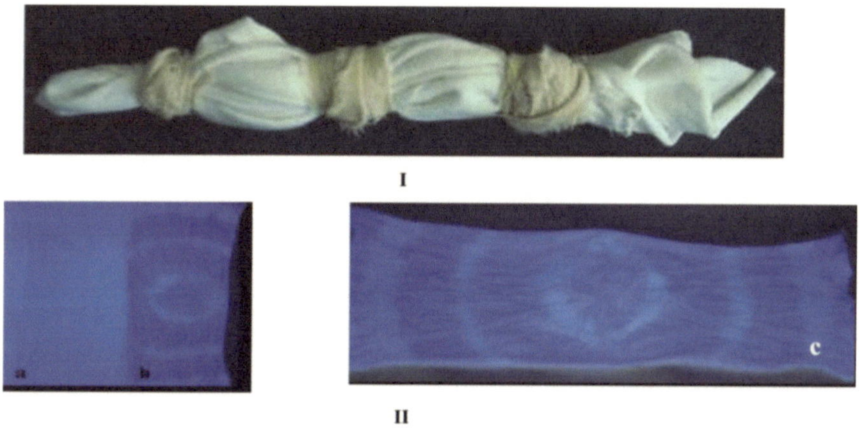

Figure 4.
The photographs of fabrics treated by developed method (I: optical bleached fabric is tied in knots, tightly bound with thread; II: a = optical bleached fabric before ozonation; II: b/c = optical bleached fabric after ozonation in accordance with developed method [40]). (Thanks to Textile and Apparel for copyright).

Gashti et al. studied on surface oxidation of cellulose by ozone gas. The aim of the study is to investigate the influence of ozonation on the performance of the fluorocarbon monomer on cotton. As a result of the study, fluorocarbon efficiency on cotton is remarkably improved by ozonation before fluoromonomer grafting. The contact angle tests and microscopic appearances show that contact angle increases because of the higher efficiency of the water repellent polymer on the treated cotton by ozone [41].

In Bahtiyari and Benli's study, ozone ultrasound humidifier combines to bleach the cotton fabric before dyeing with green walnut shells. As a result, treated cotton with ozone can be dyed with green walnut shells, and the colors of the natural dyed fabrics are good. Even if no mordanting agent is used in natural dyeing, the fastnesses are sufficient [42].

Bahtiyari and Benli proposed a green process line in their paper. In their study, cotton fabrics are treated by ozone gas and ultrasound before natural dyeing without mordant (**Figure 5**). Natural dyes are nutshell, orange tree leaves, and alkanet roots. Finally, ozone and ultrasound are used for the pretreatment of cotton before natural dyeing without mordanting agent. At the same time, fastnesses of all the dyed samples are generally sufficient, except light fastness. But light fastness of dyed samples with pomegranate peels is only sufficient [43].

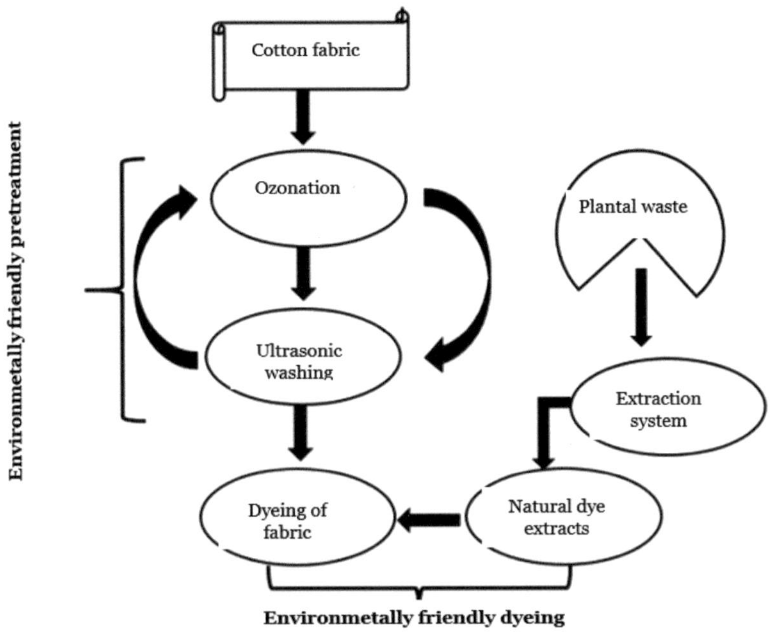

Figure 5.
Green process line [43].

Erdem and Bahtiyari combined ultrasound and ozone during the pretreatment of cotton slivers. Ultrasound is used in scouring process, and as expected, ozone is for bleaching process. As a result of the study, hydrophilicity of the cotton is achieved by pectinase enzyme/ultrasound combination. Meanwhile, bioscoured cotton is dyed by using pomegranate peel and green tea. In the end of the study, they suggest their process line for production of medical and cosmetic textiles [44].

Benli and Bahtiyari applied ozone to natural dyed cotton fabrics in their study. Aim of the study is to establish an alternative natural dyeing method without mordanting agents. Ozonation ways are given below:

- After the dyeing and washing processes, application of ozone gas to dyed fabrics, which are wet.
- Before washing process, application of fresh and cold water through ozone gas to dyed fabrics.
- After dyeing, application of fresh and cold water through ozone gas to dyed fabrics. Then, washing of the treated fabrics.

Different ozone application ways present various shades and effects due to chemical structure of natural dyes. Different types of fastnesses are examined in terms of ozone application ways and various mordanting agents. Generally, all dyed samples have high fastness except for light fastnesses. Direct ozonation of wet dyed samples improves the rubbing fastness values [45].

Kan et al. examined the effect of plasma-induced ozone treatment on the color fading of reactive dyed cotton fabric. According to the results, color fading effect is increased by increasing ozonation time, and air ratio has considerable effect on color fading. Color levelness of the ozone-treated fabrics is excellent [5, 46].

Eren et al. studied the color stripping of reactive dyed cotton by ozonation. The parameters are ozonation time and type of reactive dyes. The longest ozonation time gives the best color stripping result. COD value of effluent from ozonation is less than that of conventional reductive treatment [47].

Yiğit et al. discussed ozonation for discharge printing of reactive dyed cotton in their paper. The aim of the study is to use ozone gas instead of reductive agent and caustic soda in discharge printing. Color discharge increases at higher gas flow rates and prolonged ozonation times. According to results, ozone gas can be used for discharge printing. However, contour sharpness of conventional discharge printing is much better than that of ozonation. It is not as excellent as contour sharpness of conventional discharge printing [48].

Zhong et al. investigated color-fading process of sulfur-dyed cotton fabric by a plasma-induced ozone. As the results, the plasma-induced ozone color-fading treatment can be used to remove the color from the dyed fabric and the effect is uniform and even [49].

Perinçek et al. studied the ozonation of jute. The results indicate that the ozonation conditions for the best whiteness degree are fabric at pH 7, 60% WPV, and temperature of 23–25°C. The lignin content and DP values of fabrics are reduced by ozonation [50].

Perinçek et al. also combined ozonation and hydrogen peroxide bleaching in their paper. Linen fabrics are bleached in two steps. First, the linen fabric is treated by ozone. Then, it is bleached by hydrogen peroxide. The treatment conditions are optimized statistically [51].

Kurban et al. examined the ozonation of nettle biofiber in their study. Different bleaching methods are applied to nettle fiber fabric. They are:

- ozonation,
- ozonation in the presence of ultrasonic homogenizer,
- combination of the conventional bleaching and ozonation process, and
- combination of laccase enzyme and ozonation process.

As a result of the study, ozonation improves the whiteness of nettle fiber fabric. Among all the bleaching ways, the highest whiteness is obtained from combination of the hydrogen peroxide bleaching and ozonation process [52].

3.3 Ozonation of protein fibers

One of the early studies is about ozonation of wool garments. The aim of the study is to obtain shrink-resisted wool garments and fabric. Therefore, a continuous or batch treater was designed, and wool fabric and garments hung in cabin containing ozone. It was found out that circulation of the vapor around the garments and fabric is inevitable for rapid reaction. Fabric construction is very important for desired degree of shrink resistance. It is claimed that ozone-steam process is a solution to the felting problem of wool [53].

Rahmatinejad et al. discussed innovative hybrid fluorocarbon coating on wool treated with UV/ozone. The application of fluorocarbons on the wool fabrics is a problem because of chemistry and structure of the fiber surface. Therefore, UV/ozone treatment is proposed as a solution to this problem. Firstly, wool is modified by UV/ozone treatment. Hydrophilicity of treated wool is remarkably better than untreated wool. UV/ozone treatment can be applied to one side of the fabric and hybrid functional fabrics with two different properties on each side of the fabric are thus obtained [54].

The chlorine/Hercosett process is the most common treatment for the wool dyeing. It causes environmental problems because of the pollution of wastewater with absorbable organic halides (AOX). Therefore, ozonation is an alternative surface modification method for improving wool dyeability [6].

Micheal and El-Zaher examined the effect of ultraviolet and ozone combination for different times on wool. As a result of the study, wetting of the wool is improved by the ultraviolet/ozone process because of surface modification. It means that there is an increase in amorphous areas of the treated wool. Ultraviolet/ozone oxidizes cystine bond on the surface of the wool fabrics and generates free radical species. They support dye uptake [6].

Shao et al. investigated the effect of UV/ozone exposure and peroxide pad-batch bleaching on the printability of wool. It is found that peroxide pad-batch bleaching can prevent the yellowness caused by UV/ozone treatment and improved the wettability of the treated wool in a short time. Printability performance of treated wool is similar to that of chlorinated wool [55].

Sargunamani and Selvakumar investigated the effects of ozonation on raw and degummed tassar silk fabrics. Ozone treatment is compared with soap degumming and hydrogen peroxide treatment. Soap treatment of silk is less harsh than ozonation. Peroxide treatment causes lower yellowing index compared to ozonation [6, 56].

Balcı et al. discussed the effects of plasma and ozone treatments on silk in their paper. In this study, raw and degummed silk fabrics are treated with low-frequency oxygen plasma and ozone. The processes are applied to the fabrics individually and alternately. According to the results, fabrics treated with ozone have more yellowing index that of plasma treatment. Increasing the treatment time of plasma and ozonation processes causes increase in yellowness and decrease in whiteness [57].

In this study, Perinçek et al. examined role of the fiber moisture, pH, and treatment time during ozonation on the dyeing properties of Angora rabbit. It is observed that ozonation increases the whiteness degree and dyeability property of the fibers. Ozone oxidizes cysteine linkage in the surface of fiber to cysteic acid [6, 58].

Perinçek et al. combine ozone and ultrasound in their paper. First, Angora rabbit fibers are bleached by ozonation. Then, treated fibers are dyed by the aid of ultrasound. It is indicated that the ozonation and ultrasonic dyeing improves the dyeability of Angora rabbit fiber considerably [59].

Atav and Yurdakul investigated the ozonation of mohair fibers. The optimum conditions of ozonation process are W.P.V. 60%, pH 7, and 30 min. Dyeability of the mohair fibers is improved by ozonation [6].

Perinçek et al. discussed bleaching of soybean fabric by different treatments combined with ozonation in their paper. Combined process is ozonation, oxidative, and reductive bleaching. Process steps are shown in **Figure 6**.

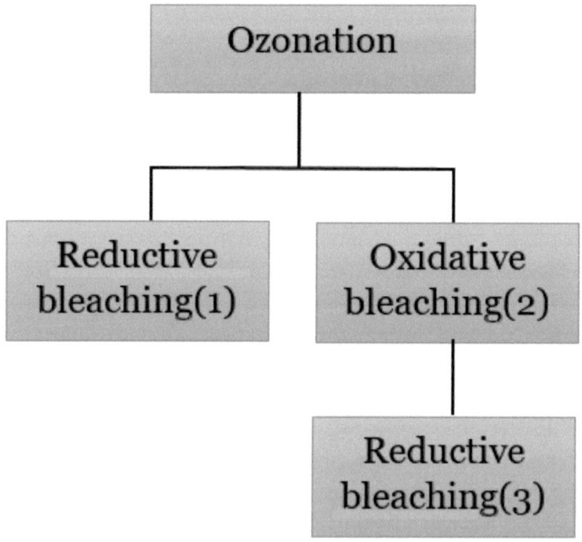

Figure 6.
Steps of bleaching process [60] (1: ozonation + reductive bleaching, 2: ozonation + oxidative bleaching, 3: ozonation + oxidative bleaching + reductive bleaching).

Consequently, combined bleaching processes improve whiteness and hydrophilicity degree, wettability of the soybean fabrics significantly [60].

Benli and Bahtiyari examined dyeing of casein fibers with natural dye after ozonation. Casein fabrics are bleached by ozone. However, whiteness degrees of the treated fabrics are limited [61].

3.4 Ozonation of the other fibers

Hydrophilicity of synthetic polymer surfaces can be accomplished by ozone. The reactive molecules on 3-D structures are covered uniformly during the ozonation. Ozone treats not only the surface and penetrates through the polymer bulk. Ozone self-decomposes rapidly in water producing free radicals, a stronger oxidant than ozone itself. This property was utilized to produce hydrophilic and highly reactive high-density polyethylene (HDPE) films [62].

Yang et al. studied the effect of ozone on aramid fibers. They found out that surface morphology of aramid fabrics does not have obvious change after the treatment. Wicking effect increases slightly with increasing ozonation time. Ozonation treatment does not have significant effect on the tenacity and elongation of the fibers. However, the tenacity and elongation of aramid yarns improve significantly after ozonation and increase with increasing ozonation time. The authors claim that ozonation process extracts foreign matters from the surface of the fiber and establishes oxygen-containing functional groups. The importance of oxygen-containing functional groups is to support adhesion to the matrix [63].

Rahmatinejad et al. investigated enhancement in polyester materials' hydrophobicity by surface modification via chemical pretreatment, UV/ozone irradiation, and fluorocarbon finishing combinations. The study concentrates on the application of UV/ozone radiation together with various chemical pretreatments on fabrics and their effects on the fluorocarbon finishing performance. In surface modification, UV/ozone irradiation prior to fluorocarbon treatment results in more hydrophobic polyester fiber surface than only fluorocarbon-treated fabric. Because of erosion, redeposition, and the melting effects of UV/ozone-irradiation, proper unevenness of the fiber surface is formed by UV/ozone radiation [64].

Elnagar et al. studied dyeability of polyester and nylon fabrics treated with UV/ozone radiation. Mordant is ferrous sulfate. Natural dyes are curcumin and saffron dyes. The results show that dyeability of both fabrics with curcumin and saffron natural dyes increases with aid of UV/ozone [65, 66].

Atav and Namırtı investigated effect of ozonation process on dyeing of polyamide fabrics with walnut rind natural dye. Color yield is increased by ozonation before dyeing polyamide fabric, and ozone gas does not affect the color nuance and fastness properties negatively [67].

Lee et al. discussed ozone-gas treatment of nylon 6 and polyester fabrics in their paper. It is appeared that the O1s relative intensity increases for nylon 6 and polyester fabrics. Oxygen is included in the form of OCOH and OCOOH. Therefore, hydrophilicity of treated fabrics is higher than untreated ones. Ozonation changes the crystalline and amorphous regions, especially for polyester fiber. Moisture regain, water absorption, and dyeing properties increase despite an increase in the crystallinity. On the other hand, ozonation affects the brittle hand of the fabric [68].

Lee et al. studied ozonation of cationic dyeable polyester and poly(butylene terephthalate) fibers too. Although water absorption is improved by ozonation, crystallinity index increases a little bit. They claim that ozonation changes fiber surface. On the other hand, internal structure of both fibers is also changed by the treatment. Therefore, it has effect on the dyeing properties of the fibers. After ozonation, dyeing rate with the cationic dye increases exceptionally. However, increasing dyeing rate with disperse dye is not so significant [69].

Eren and Aniş suggest ozone treatment of polyethylene terephthalate fibers after dyeing as a novel after-clearing method. Results indicate that the trimer removal rates of ozone treatment are quite similar to the conventional reduction clearing for 1-min and higher for 3-min ozone treatments. The treatment time at 130°C is also efficient on the amount of surface trimer [70].

Eren studied on combination of after-clearing and decolorization by ozonation after disperse dyeing of polyester. He claims that encouraging results from decolorization and wash fastness tests are obtained with a 3-min ozonation period in the dyebath at room temperature. Decolorization and COD removal ratios are up to 67 and 62%, respectively [71].

Eren et al. discussed after-clearing of disperse dyed polyester with gaseous ozone in their paper. They propose that a new ozonation method is adopted to continuous treatment lines. Proposed method is different from exhaust application method in early papers [71, 72]. Ozone gas from the generator blasts through the wet fabric. Depending on the type of the disperse dyes, ozonation time is different for wash fastness results which is comparable to that of conventional reduction clearing method. According to results from tensile strength tests and scanning electron microscopy analysis, ozonation does not cause any serious damage to the fabrics [73].

3.5 Advantages and disadvantages of use of ozone in the textile industry

Wet processing of textile materials consumes large amounts of electricity, fuel, and water. Therefore, greenhouse gas emissions and contaminated effluent are environmental problem. Ozone treatment proposed a solution to environmental pollution from textile wet processes [5, 74, 75]. Use of ozone in the textile industry has advantages and limitations [4].

Advantages of ozonation in the textile industry [4, 5, 13, 74]:

- lower water and chemical consumption and time loss of ozonation process than conventional wet processes,
- no need to store chemicals compared to the other conventional methods,
- no dangerous waste because of decomposition of ozone into oxygen,
- no halogenated organic compounds (AOX) in waste water,
- combination with novel technologies like UV, plasma, and ultrasound,
- different pattern and fading effects soon in denim washing,
- improving dyeability of fibers,
- more ecological antifelting treatment than conventional methods,
- higher whiteness than conventional bleaching processes, and
- treatment of hygienic nondurable products like sheets, gauze bandage, tissues, bib, hydrophilic cotton, etc., due to disinfectant property of ozone.

Limitations of ozone treatment [4, 5, 13, 14, 32, 74, 76, 77]:

- prevention of yellowing problem with after-treatment like catalase treatment, reductive washing, etc.
- high strength loss in textile materials due to illiterate use of ozone,
- difficulty in using ozone gas due to suitability of textile finishing machine for wet processes,
- except for stainless steel, corrosion in metal parts of finishing machine due to high oxidation potential of ozone,
- damage possibility of plastics of the finishing machine due to high oxidation potential of ozone,
- high capital investment for new machinery setups,
- necessity of onsite generation because of unsuitable for storage,
- unsuitable for storage due to decomposition of ozone quickly,
- use of ozone in illiterate way due to occupational health and safety,
- needs regular monitoring and alarm system in the mill for any leakages, and
- flammability and explosivity of ozone.

Finally, ozonation within a closed system can be called as environmental process [4]. However, limitations of the ozone have not to be forgotten by users.

Author details

Ayşegül Körlü
Textile Engineering Department, Faculty of Engineering, Ege University,
Bornova-İzmir, Turkey

*Address all correspondence to: aysegulkorlu@gmail.com

References

[1] Broadbent AD. Basic Principles of Textile Coloration. West Yorkshire: Society of Dyers and Colourists; 2001. 29 p. ISBN: 0 901956 76 7

[2] Muthu SS, editor. Sustainable Fibres and Textiles. Duxford: Woodhead Publishing; 2017. 29 p. ISBN: 978-0-08-102042-5

[3] Perinçek S, Bahtiyari İ, Körlü A, Duran K. New Techniques in Cotton Finishing Year. Textile Research Journal; 2009;**79**(2):121-128. DOI: 10.1177/0040517508092022

[4] Duran K, Körlü AE, Perinçek S, Bahtiyari Mİ. Ozone: New tendency in textile finishing. In: 21st IFATCC International Congress Barselona. 2008. ISBN 13: 978-84-612-2641-2

[5] Muthu SS, editor. Sustainable Innovations in Textile Chemical Processes. Singapore: Springer Nature Singapore Pte Ltd.; 2018. 11 p. ISBN: 978-981-10-8491-1

[6] Günay M, editor. Eco-Friendly Textile Dyeing and Finishing. England: Intech; 2013. 103 p. DOI: 10.5772/3436

[7] Tabar IB. Ozone as oxidant for biomass pretreatment and nanocellulose production [thesis]. Indiana: The Purdue University; 2017

[8] https://dc.edu.au/dot-point-summary-chemical-monitoring-management/#25_Describe_ozone_as_a_molecule_able_to_act_both_as_an_upper_atmosphere_UV_radiation_shield_and_a_lower_atmosphere_pollutant

[9] Bocci V, Borrelli E, Travagli V, Zanardi I. The ozone paradox: Ozone is a strong oxidant as well as a medical drug. Medicinal Research Reviews. 2009;**29**(4):646-682. DOI: 10.1002/med.20150

[10] Perinçek DS, Duran K, Körlü AE, Bahtiyari Mİ. An investigation in the use of ozone gas in the bleaching of cotton. Ozone: Science and Engineering. **29**:325-333. DOI: 10.1080/01919510701509578

[11] https://saylordotorg.github.io/text_general-chemistry-principles-patterns-and-applications-v1.0/s13-molecular-geometry-and-covalen.html

[12] Iglesias SC. Degradation and biodegradability enhancement of nitrobenzene and 2,4-dichlorophenol by means of advanced oxidation processes based on ozone [thesis]. Barcelona: Universitat De Barcelona; 2002

[13] Prabha V, Barma RD, Singh R, Madan A. Ozone technology in food processing: A review. Trends in Biosciences. 2015;**8**(16):4031-4047. ISSN: 0974-8

[14] Wojtowicz JA. Ozone Kirk-Othmer Encyclopedia of Chemical Technology. 2005;**17**:768. DOI: 10.1002/0471238961.1526151423151020.a01.pub2

[15] Steinberg M, Beller M, Powell JR. Thu. Large scale ozone production in chemonuclear reactors for water treatment. United States. DOI: https://www.osti.gov/servlets/purl/4332197

[16] Cesarino I, Moraes FC, Machado SAS, Filho JP, Cardoso AA. A new indirect electrochemical method for determination of ozone in water using multiwalled carbon nanotubes. Electroanalysis. 2011;**23**(6):1512-1517. DOI: 10.1002/elan.201000756

[17] Bader H, Hoigni J. Determination of ozone in water by The Indigo Method. Water Research. 1981;**15**:449-456

[18] Tomiyasu H, Gordon G. Colorimetric determination of ozone in water based on reaction with bis(terpyridine)iron(II). Analytical

Chemistry. 1984;**56**:752-754. DOI: 10.1021/ac00268a036

[19] https://www.chemetrics.com/image/data/product/pdf/Measurement_of_Ozone_White_Paper_Final.pdf

[20] Tjahjanto RT, Dhesy GR, Wardani S. Ozone determination: A comparison of quantitative analysis methods. Journal of Pure and Applied Chemistry Research. 2012;**1**(1):18-25. DOI: 10.21776/ub.jpacr.2012.001.01.103

[21] Grunwell J, Benga J, Cohen H, Gordon G. A detailed comparison of analytical methods for residual ozone measurement. Ozone: Science & Engineering: The Journal of the International Ozone Association. 1983;**5**(4):203-223. DOI: 10.1080/01919518308550989

[22] Sullivan DE, Hall LC, Ambrosi MD, Roth JA. Analysis of ozone in aqueous solutions using a modified iodometric technique with as (III). Ozone: Science & Engineering: The Journal of the International Ozone Association. 1980;**2**(2):183-193. DOI: 10.1080/01919518008550878

[23] Delgado M. Ozone concentration measurements. State of the art. Revista Española de Ozonoterapia. 2011;**1**(1):87-92

[24] Majewski J. Methods for measuring ozone concentration in ozone-treated water. Przegląd Elektrotechniczny. 2012;**88**(9b):253. ISSN: 0033-2097

[25] Liptak BG, editor. Instrument Engineers' Handbook. 4th ed. Vol. 1. Florida: CRC Press. 1540 p. ISBN 0-8493-1083-0

[26] Yu X, Yang H, Sun L. Determination of chemical oxygen demand using UV/O_3. Water, Air, and Soil Pollution. 2016;**227**:458. DOI: 10.1007/s11270-016-3154-y

[27] Muthu SS, editor. Sustainable Innovations in Apparel Production. Singapore: Springer Nature Singapore Pte Ltd; 2018. DOI: 10.1007/978-981-10-8591-8_2

[28] Özdemir D, Duran K, Bahtiyari Mİ, Perinçek S, Körlü AE. Ozone bleaching of Denim fabrics. American Association of Textile Chemists and Colorists. 2008;**8**(9):40-44. ISSN: 1532-8813

[29] Khalil E. Sustainable and ecological finishing technology for Denim jeans. AASCIT Communications. 2015;**2**(5):159 p. ISSN: 2375-3803

[30] Körlü AE, Yapar S, Perinçek S, Yılmaz H, Bağıran C. Dye removal from textile waste water through the adsorption by pumice used in stone washing. AUTEX Research Journal. 2015;**15**(3):158 p. DOI: 10.1515/aut-2015-0012

[31] Paul R editor. Denim Manufacture, Finishing and Applications. United Kingdom: Woodhead Publishing; 2015. 313 p. 405 p. 425 p. DOI: 10.1016/B978-0-85709-843-6.00011-1, DOI: 10.1016/B978-0-85709- 843-6.00013-5, DOI: 10.1016/ B978-0-85709-843-6.00014-7

[32] Muthu SS editor. Sustainability in Denim. United Kingdom: Woodhead Publishing; 2017. 176 p. 330 p. DOI: 10.1016/B978-0-08-102043-2.00007-1, DOI: 10.1016/B978-0-08-102043-2.00012-5

[33] Hmida SB, Ladhari N. Study of parameters affecting dry and wet ozone bleaching of denim fabric. Ozone: Science & Engineering. 2016;**38**(3):175-180. DOI: 10.1080/01919512.2015.1113380

[34] Bağıran İC, Duran K, Körlü A. Comparison with use of ozone and other methods in bleaching of denim and environmental advantages of ozone. 15th recent progress symposium on textile technology and chemistry. 14-16 May 2015. Bursa

[35] Beşen BS, Balcı O. Fading of cotton yarn colored with C.I. vat blue I (indigo dye) via ozone application. Ozone: Science & Engineering. 2016;**38**(5):395-409. DOI: 10.1080/01919512.2016.1204529

[36] He Z, Li M, Zuo D, Yi C. The effect of denim color fading ozonation on yarns. Ozone: Science & Engineering. 2018;**40**(5):377-384. DOI: 10.1080/01919512.2018.1435259

[37] Eren HA, Öztürk D. The evaluation of ozonation as an environmentally friendly alternative for cotton preparation. Textile Research Journal;**81**(5):512-519. DOI: 10.1177/0040517510380782

[38] Maqsood HS, Wiener J, Baheti V, Eldeeb M. Ozonation: A green source for oxidized cotton. Fibres & Textiles in Eastern Europe. 2016;**24**(1(115)):19-21. DOI: 10.5604/12303666.1168523

[39] Turhan Y, Soydaş Ş. The effects of ozone bleaching and ozone desizing method on whiteness and water absorption of 100% cotton terry fabrics. International Journal of Materials Science and Applications. 2018;7(3):85-94. DOI: 10.11648/j.ijmsa.20180703.13

[40] Perinçek S. Ozonation: A new patterning method for optical bleached fabrics. Tekstil ve Konfeksiyon. 2016;**26**(1):76-83. ISSN: 1300-3356

[41] Gashti MP, Pournaserani A, Ehsani H, Gashti M. Surface oxidation of cellulose by ozone-gas in a vacuum cylinder to improve the functionality of fluoromonomer. Vacuum. 2013;**91**:7-13. DOI: 7e13 DOI10.1016/j.vacuum.2012.10.015

[42] Bahtiyari Mİ, Benli H. Ozone bleaching of cotton fabrics with the aid of ultrasonic humidifier. Cellulose. 2016;**23**:2715-2725. DOI: 10.1007/s10570-016-0978-y

[43] Benli H, Bahtiyari Mİ. Combination of ozone and ultrasound in pretreatment of cotton fabrics prior to natural dyeing. Journal of Cleaner Production. 2015;**89**:116-124. DOI: 10.1016/j.jclepro.2014.11.007

[44] Erdem A, Bahtiyari MI. Ultrasonic-bioscouring and ozone based bleaching of cotton slivers and coloration of them with natural dye sources. Journal of Cleaner Production. 2018;**188**:670-677. DOI: 10.1016/j.jclepro.2018.03.166 0959-6526

[45] Benli H, Bahtiyari Mİ. Combination of dyeing method and ozone after-treatment to apply natural dyes on to cotton fabrics. Ozone: Science & Engineering. 2018;**40**(1):44-53. DOI: DOI10.1080/01919512.2017.1336926

[46] Kan C, Cheun HG, Chan Q. A study of plasma-induced ozone treatment on the colour fading of dyed cotton. Journal of Cleaner Production. 2016;**112**:3514-3524. DOI: 10.1016/j.jclepro.2015.10.100

[47] Eren S, Gümüş B, Eren HA. Colour stripping of reactive-dyed cotton by ozone treatment. Coloration Technology; 2016;**132**:466-471. DOI: 10.1111/cote.12240

[48] Yiğit İ, Eren S, Eren HA. Ozone utilisation for discharge printing of reactive dyed cotton. Coloration Technology; 2017;**134**:13-23. DOI: 10.1111/cote.12306

[49] Zhong D, Liu Y, Cheung N, Kan C, Chua HA. Parameter study of the effect of a plasma-induced ozone colour-fading process on sulphur-dyed cotton fabric. Processes. 2018;**6**:81. DOI: 10.3390/pr6070081

[50] Perinçek S, Bahtiyari Mİ, Körlü ve AE, Duran K. Ozone bleaching of jute fabrics. AATCC Review. 2007;7(3):34-39

[51] Perinçek S, Duran K, Korlu AE. Combination of ozonation and hydrogen peroxide bleaching for linen

fabrics: Optimization of the process using experimental design technique. Ozone: Science & Engineering: The Journal of the International Ozone Association. 2013;**35**(4):316-327. DOI: 10.1080/01919512.2013.788979

[52] Kurban M, Yavas A, Avinç O, Eren HA. Nettle biofibre bleaching with ozonation. Industria Textila. 2016;**67**(1):46-54. ISSN: 1222-5347

[53] Thorsen J, Sharp D L, Randall V G. Vapor-phase ozone treatment of wool garments. Textile Research Journal. 1979(4 April);**49**:190-197

[54] Rahmatinejad J, Khoddami A, Avinc O. Innovative hybrid fluorocarbon coating on UV/ozone surface modified wool substrate. Fibers and Polymers. 2015;**16**(11):2416-2425. DOI: 10.1007/s12221-015-5120-5

[55] Shao J, Liua J, Carr CM. Investigation into the synergistic effect between UV/ozone exposure and peroxide pad–batch bleaching on the printability of wool. Coloration Technology. 2001;**117**:1472-3581

[56] Sargunamani D, Selvakumar N. Effects of ozone treatment on the properties of raw and degummed tassar silk fabrics. Journal of Applied Polymer Science. 2007;**104**:147-155. DOI: 10.1002/app.24761

[57] Balcı N, Ömeroğullari Z, Kut D, Eren HA. Effects of plasma and ozone treatments. On tensile and whiteness properties of 100% silk. Uludağ University Journal of The Faculty of Engineering. 2015;**2**:43-56. DOI: 10.17482/uujfe.33090

[58] Perinçek S, Bahtiyari MI, Korlu AE, Duran K. Ozone treatment of Angora rabbit fiber. Journal of Cleaner Production. 2008;**16**:1900-1906. DOI: 10.1016/j.jclepro.2008.01.005

[59] Perinçek S, Bahtiyari MI, Korlu AE, Duran K. Effect of ozone and ultrasound on the fiber properties of angora rabbit. Journal of Applied Polymer Science. 2011;**120**:3119-3125. DOI: 10.1002/app.28779

[60] Perinçek S, Duran K, Korlu AE. The bleaching of soybean fabric by different treatments combined with ozonation. Ozone: Science & Engineering. 2015;**37**(3):195-202. DOI: 10.1080/01919512.2014.956860

[61] Benli H, Bahtiyari Mİ. Dyeing of casein fibers with onion skin-based natural dye sources after ozonation. Ozone: Science & Engineering. 2018;**40**(2):141-147. DOI: 10.1080/01919512.2017.1341300

[62] Hauser P, editor. Textile Dyeing. England: Intech; 2011. 268 p. DOI: 10.5772/800. https://www.intechopen.com/books/textile-dyeing

[63] Wang Y, Wiener J, Militky J, Mishra R, Zhu G. Ozone effect on the properties of aramid fabric. AUTEX Research Journal. 2017;**17**(2):164-169. DOI: 10.1515/aut-2016-0027

[64] Rahmatinejad J, Khoddami A, Mazrouei-Sebdani Z, Avinc O. Polyester hydrophobicity enhancement via UV-ozone irradiation, chemical pre-treatment and fluorocarbon finishing combination. Progress in Organic Coatings. 2016;**101**:51-58. DOI: DOI 10.1016/j.porgcoat.2016.07.017

[65] Purwar S. Application of natural dye on synthetic fabrics: A review. International Journal of Home Science. 2016;**2**(2):283-287. ISSN: 2395-7476

[66] Elnagar K, Elmaaty TA, Raouf S. Dyeing of polyester and polyamide synthetic fabrics with natural dyes using ecofriendly technique. Journal of Textiles; 2014;**2014**:1. DOI: 10.1155/2014/363079

[67] Atav R, Namırtı O. Effect of ozonation process on dyeing of polyamide fabrics with a natural dye: Walnut rind. Industria Textila. 2016;**67**(4):233-237. ISSN: 1222-5347

[68] Lee M, Lee MS, Wakida T, Tokuyama T, Inoue G, Ishida S, et al. Chemical modification of nylon 6 and polyester fabrics by ozone-gas treatment. Journal of Applied Polymer Science. 2006;**100**:1344-1348. DOI: 10.1002/app.23382

[69] Lee MS, Lee M, Wakida T, Saito M, Yamashiro T, Nishi K, et al. Ozone-Gas Treatment of Cationic Dyeable Polyester and Poly(butylene terephthalate) Fibers. Journal of Applied Polymer Science. 2007;**104**:2423-2429. DOI: 10.1002/app.25978

[70] Eren H A, Anis P. Surface trimer removal of polyester fibers by ozone treatment. Textile Research Journal. 2009;**79**(7):652-656. DOI: 10.1177/0040517508097521

[71] Eren HA. Simultaneous afterclearing and decolorisation by ozonation after disperse dyeing of polyester. Coloration Technology. 2007;**123**:224-229. DOI: 10.1111/j.1478-4408.2007.00087.x

[72] Eren HA. Afterclearing by ozonation: a novel approach for disperse dyeing of polyester. Coloration Technology. 2006;**122**:329-333. DOI: 10.1111/j.1478-4408.2006.00049.x

[73] Eren HA, Ozturk D, Eren S. Afterclearing of disperse dyed polyester with gaseous ozone. Coloration Technology. 2012;**128**:75-81. DOI: 10.1111/j.1478-4408.2011.00350.x

[74] Hasanbeigi A, Price L. A technical review of emerging technologies for energy and water efficiency and pollution reduction in the textile industry. Journal of Cleaner Production. 2015;**95**:30-44. DOI: 10.1016/j.jclepro.2015.02.079

[75] Muthu SS editor. Handbook of Sustainable Apparel Production. Boca Raton: CRC Press. 2015. 386 p. ISBN 13: 978-1-4822-9939-7

[76] Perinçek S, Bahtiyari Mİ, Duran K, Körlü AE. Yellowing tendency of ozonated cotton fabric and ways to prevent this undesirable side effect. Journal of the Textile Institute. 2009;**100**(8):738-746. DOI: 10.1080/00405000802273608

[77] Hussain T, Wahab A. A critical review of the current water conservation practices in textile wet processing. Journal of Cleaner Production. 2018;**198**:806-819. DOI: 10.1016/j.jclepro.2018.07.051 0959-6526

Chapter 3

Chemical and Tinctorial Aspects Related to the Reuse of Effluents Treated by Ozonation in Dyeing Processes

Pablo Colindres Bonilla

Abstract

The purpose of this chapter is to study the interactions that are established between inorganic auxiliaries and the by-products of contaminants present in effluents coming from dyeing operations during oxidation treatment processes using ozone, and the influence of auxiliaries and by-products on the behavior of dyes in subsequent dyeing processes using the treated water. Carrying the treatment until the complete elimination of the contaminants present in it is a very expensive operation. Because of this, it is chosen to discolor and reuse the spent dyebaths treated as many times as possible to take advantage of the water and the inorganic salts contained therein. The variable composition of the dyebaths involves kinetic aspects during the treatment, which is important to take into account in the design of the process. Various by-products are already generated from the beginning of the treatment, which will have an influence on the following stages of the same treatment process, as well as on the kinetics of the dyeing processes carried out using the treated water and on the results obtained in such dyeing processes. All this will depend on the chemical and dyeing class to which the dyes used during the subsequent dyeing processes belong.

Keywords: water reuse, oxidation, ozone, textile, dyes

1. Introduction

In order to demonstrate the influence of the mentioned auxiliaries and by-products on the dyeing processes carried out using treated water, this chapter is focused on the discoloration of aqueous solutions of two reactive dyes of wide application in the textile industry, Reactive Blue 160 (RB160) and Reactive Red 141 (RR141), through the application of ozone, with the purpose of reusing treated water in dyeing processes of cotton goods, taking advantage of the large quantities of inorganic salts contained in a spent dyebath.

The greatest water consumption in a textile plant occurs in the finishing areas, since the processes carried out there use substances in aqueous solution to modify the characteristics of the processed articles, being the main one of these characteristics the coloration. Reactive type dyes have been used since 1956 for the dyeing of articles that have a high resistance to deterioration by conventional washing, since they are compounds that react to form covalent bond dye fiber [1, 2]. This type of resistance is known as washing fastness. However, reactive dye groups that do not react with the textile substrate are hydrolyzed during the dyeing process, which, in addition to drastically reducing the dye's performance during dyeing and increasing the amount of residual dye, precludes its reuse by reconstitution of the dyeing bath and causes them to be discarded directly to the effluent. A part of the dye (up to 50%) is hydrolyzed during dyeing. This dye is an important source of contamination and makes the reuse of water impossible [3].

Within the category of reactive dyes, the azoic type represents more than 95% of the existing structures in the most used color shades. In dyes of this kind, the color originates from nitrogen-nitrogen double bonds or azo bonds (—N=N—). In general terms, this is the chromophore group (color generator) most frequently used. The main field of application of reactive dyes is the dyeing of articles made of textile fibers derived from cellulose. Its application on woolen articles is more limited, since this last fiber presents good washing fastness when it is dyed with another type of higher-performance dyes, such as, for example, acid dyes [4].

For the elimination of dyes present in wastewater from the textile industry, various processes have been used, both physical and chemical, as well as biological processes. However, reactive dyes are toxic compounds that are difficult to eliminate by biodegradation or separation methods, such as membrane filtration, coagulation-flocculation, or adsorption on activated carbon. The elimination of the color of a wastewater of this type can be carried out to comply with the environmental norms in the discharges, or else, in order to reuse the water. Both in the case of the treatment with activated carbon and in the coagulation-flocculation, results of color removal are obtained just over 50%, which are insufficient for any of the mentioned objectives [5].

Regarding the elimination of color to reuse water, very remarkable results have been achieved, close to 100%, through the use of nanofiltration membranes. However, this method does not solve the problem; it only separates the contaminant, without eliminating it. On the other hand, it presents technical difficulties, in terms of the difference in pressure due to the occlusion of the membranes, which causes its deterioration, in addition to the disposition of the separated contaminant. In the same way, chemical methods, particularly oxidation, have been used to eliminate the dyes present in wastewater in order to reuse it. The best results in terms of complete elimination of the dye until its mineralization have been obtained by photocatalysis, either homogeneous (photo-Fenton processes) or heterogeneous (using semiconductors, especially TiO_2) [6].

The ozonation, on the other hand, has been used individually, combined with hydrogen peroxide (peroxonous), as well as in conjunction with the application of UV radiation or the presence of activated carbon, and various combinations of these methods. This type of combination, as well as photocatalytic and Fenton processes, is known as advanced oxidation processes (AOPs) [6].

Ozone is an oxidant that does not generate its own waste; this allows obtaining a treated water with suitable characteristics to be used in dyeing processes of textile materials. In addition, the use of ozone individually enables a better understanding of the oxidation process, by avoiding interferences due to the contributions of the processes involved in the combined methods [7].

The dyes were selected for study because they are among the highest consumption by the textile industry. On the other hand, these compounds belong to one of the groups of dyes that are used more frequently: RB160 and RR141 are monochlorotriazinic reactive dyes, considered as moderate reactive and dyed at temperatures of 80°C [8].

In this study, the partial oxidation of the dyes is verified, since the complete mineralization of these products by ozonation is an extremely long process, which greatly increases the cost of a process of this type and, consequently, is not the most frequent form of work in real water treatment systems.

The oxidation of the dye implies that the treated water, at the end of the process, contains intermediate oxidation products. In the case of the reuse of water treated in dyeing processes, it is necessary to verify that such by-products do not interfere with the results obtained in terms of dyeing quality. For this purpose, dyeing tests were carried out in the laboratory with two of the most frequently used dye classes for dyeing cellulosic fibers: direct dyes and reactive dyes. These tests were evaluated by a standard procedure used in the production areas of the textile industry to evaluate dyeing results compared against a reference standard or sample to match.

For the dyeing tests, a common trichromy of reactive dyes was used, that is, the cotton samples were dyed using Reactive Yellow 84 (RY84), Reactive Red 141 (RR141), and Reactive Black 5 (RB5), separately. While in the case of dyeing with direct dyes, a trichromy consisting of Direct Yellow 50 (DY50), Direct Red 23 (DR23), and Direct Blue 80 (DB80) was used. This trichromy of direct dyes was chosen because, like the three reactive dyes under study, they cover a wide range of tones and also cover the three categories of direct dye classification established by the Society of Dyers and Colourists (SDC), which is the broadest usage classification [8].

The results obtained show that the discoloration of the compounds studied is rapid even at very low concentrations of ozone, as is the case of the concentration used in this study. However, the decomposition of dyes is a much longer process and, in fact, is not carried out with ozone even using it for long periods of time at low concentrations.

The initial concentration of dye has a marked influence on the reaction rate, sharply changing the kinetics of discoloration within the range of concentrations studied. Likewise, the presence of chemical auxiliaries has a more pronounced influence at low concentrations, because at high concentrations the aggregation of the dyes increases, influencing the kinetics [1].

The ozone applied to the discoloration of the compounds studied does not negatively influence the concentration of sodium ion present in the solution but rather increases it slightly. This situation is important because the sodium ion is necessary to carry out the process of dyeing the treated water.

Within a limited range of cycles, the quality of the dyeing obtained using water treated by ozonation complies with the specifications of the method used in industry, called AATCC 173-2005 Calculation of Small Color Differences for Acceptability, established by the American Association of Textile Chemists and Colorists [9].

The influence of the by-products formed from each dye during the ozonation in the subsequent dyeing depends on the category of dye used to dye the goods, in this case of mercerized cotton. It is more pronounced for dyeing with reactive dyes.

Through the treatment scheme used in this work, it is possible to eliminate the dyes contained in the wastewater for a very short time (10 min), using very low concentrations of ozone (2 mg/L). The kinetics of the process depend on the chemical constitution of the dye and on the residual water variables such as the initial concentration of dye and dyeing auxiliaries. The ozonation allows to reuse the treated water in dyeing processes with direct dyes and reactive dyes for up to four cycles, depending on the type of dye used. The by-products resulting from simple ozonation accumulate during the cycles of reuse without significantly affecting the quality of the resulting dyeing and can be converted into biodegradable substances by a longer ozonation. **Figure 1** shows the chemical structures of RB160 and **Figure 2** shows the chemical structure of RR141.

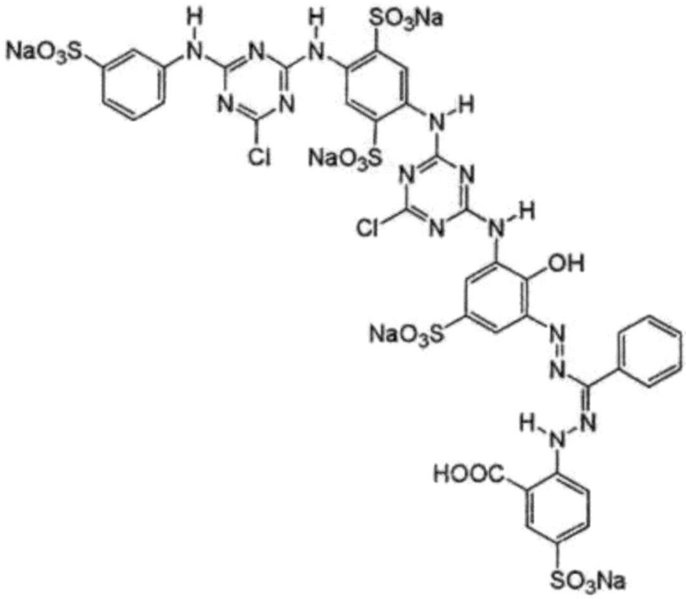

Figure 1.
Chemical structure of Reactive Blue 160 (RB160).

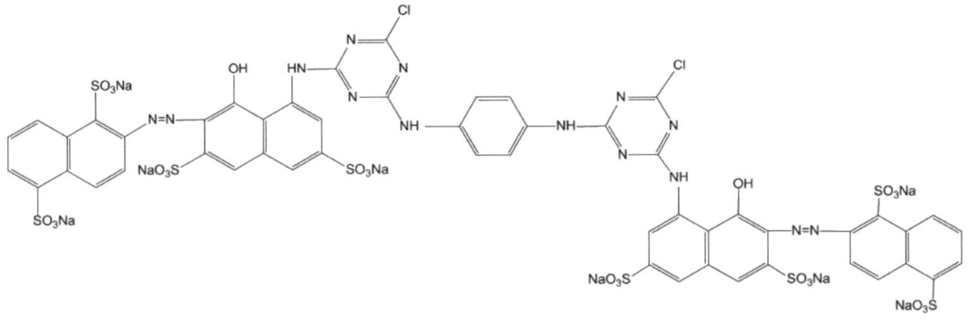

Figure 2.
Chemical structure of Reactive Red 141 (RR141).

2. Methodology

The dyes were studied individually, varying their initial concentration and varying inorganic chemical auxiliaries to simulate a dyeing effluent. Industrial samples of the three dyes were used, without any further purification. The supplier was Sinochem Ningbo, which is one of the largest presences in the Mexican market.

2.1 Ozone production

The ozone was obtained from dry air, using a corona type discharge generator (Peak Corporation). For the analysis by FTIR spectroscopy, ozone was obtained using the same generator, fed with pure oxygen (99.5%, Infra). The ozone concentration was analyzed at the output of the generator with a BMT201 (BMT Messtechnik).

2.2 Ozone reactor

The ozonation experiments were carried out in a semicontinuous glass reactor of 500 mL capacity. The reactor has an entrance in the lower part where the O_2/O_3 gaseous mixture is injected; this mixture ascends passing through a porous silica plate diffuser that evenly distributes the ozone bubbles in the liquid phase, ensuring that these are as small as possible to achieve a better mass transfer [10]. The reactor also has an outlet at the top for the ozone that does not participate in the reaction. Also, the reactor has a side key in the lower part to sample the ozone solution.

2.3 Experimental procedure

Initially, a model solution of 50 mg/L concentration was prepared for each dye. This concentration is within the range of concentrations regularly found in the effluents of the textile industry [5]. Subsequently, concentrations of 100 and 150 mg/L were used to study the effect of the concentration on the kinetics of the reaction, and solutions of 5 g/L, to carry out a preliminary analysis by FTIR.

The ozonation experiments were carried out on 400 mL of each model solution in the 500 mL reactor, with 2 mg/L of initial ozone concentration and a gaseous mixture flow (O_3/O_2) of 500 mL/min, at room temperature (20–25°C). The maximum ozonation time was 120 min. Samples were taken every 10 min and, in some cases, every 5 min.

Preparation cycles of the model solution and ozonation were made for 10 min, up to five times, with 50 mg/L solutions under the described conditions. That is, the sample solution of dye was prepared at 50 mg/L, and the ozone treatment was applied for 10 min; after this, the same sample was reconstituted with a concentration of 50 mg/L of textile dye, and it was treated again with ozone for 10 min. This cycle of contamination-discoloration was repeated up to five times with each textile dye, so that at the end there was a series of five samples of 400 mL, for each dye. The first sample had been subjected to a single cycle, the second sample to two cycles, the third sample to three, and so on until completing the five cycles. The purpose of this part of the work was to represent the recirculation of water within the process.

These samples were used for dyeing tests of textile material with reactive dyes and direct dyes. For the dyeing with reactive dyes, Reactive Black 5 (RB5), Reactive Yellow 84 (RY84), and Reactive Red (RR141) were used, while for dyes with direct dyes, a trichromy composed of Direct Yellow 50 (DY50), Direct Blue 80 (DB80), and Direct Red 23 (DR23) was selected. The dyes were carried out in a closed machine with infrared heating (Mathis). The DY50 corresponds to group A of the SDC classification of direct dyes (self-regulating); DB80, group B (controllable by means of salt); and DR23, group C (controllable by means of salt and temperature). The color difference was measured with respect to control samples, dyed with distilled water by reflectance spectrophotocolorimetry.

A second series of samples was prepared in the manner described in the previous paragraph, adding from the first cycle the equivalent to 50 mg/L of sodium ion. This was carried out in order to verify their presence throughout the ozonation cycles, using atomic absorption spectroscopy.

2.4 Analytic methods

2.4.1 pH and conductivity measurement

The variation of the pH during the ozonation was measured with a Jenway pH meter model 3310 equipment. Likewise, the variation of conductivity was measured, in order to indirectly determine the formation of ionic species. The latter was used as a criterion to verify the separation of the groups capable of forming ions, located in the molecules of the dyes, during the ozonation process.

2.4.2 UV-Vis spectroscopy

Preliminary information about the discoloration and decomposition of textile dyes in water was obtained from the variation of absorbance in the UV-Vis range, between 190 and 700 nm, with a Cary 50 (Varian) spectrophotometer. The discoloration of RR141 was analyzed at 544 nm and discoloration of RB160 at 615 nm. Similarly, the degradation of RR141 was analyzed at 288 nm, and the degradation of RB160 at 280 nm. These wavelengths correspond to the absorbance maxima of the compounds studied within the visible and UV ranges, respectively.

2.4.3 FTIR spectroscopy

An FTIR analysis was carried out to obtain information about the functional groups that are formed and disappear in the course of ozonation. For this purpose, a PerkinElmer Series 200 equipment was used. The samples of dye subjected to ozonation at intervals of 4 h, up to a maximum of 12 h, were dried and prepared in the form of a tablet mixed with KBr, to effect the measurements.

2.4.4 Atomic absorption spectroscopy

Sodium ion monitoring was carried out through the ozonation cycles of the samples prepared for dyeing, using an AA3300 spectrophotometer (PerkinElmer). The purpose of this analysis was to directly verify that ozonation did not have any adverse effect on the inorganic salt content necessary for subsequent dyeing.

2.4.5 Reflectance spectrophotocolorimetry

With the series of samples of model solutions of 50 mg/L of concentration at which the cycles of contamination with dye and subsequent ozonation were carried out up to five times, dyeing tests were made with reactive and direct dyes applied on samples of cotton, for determining the feasibility of reusing water pretreated with ozone in dyeing processes of cellulosic materials. The dyes were made on cloth prepared for this purpose. A dyeing was made with each water sample, and the difference in color kept by the samples of cloth dyed with the pretreated water was determined in relation to a reference standard, that is, a sample of the same type dyed using distilled water and maintaining the other constant conditions. The test was carried out according to the AATCC method 173-2005 of the American Association of Textile Chemists and Colorists (AATCC) called Calculation of Small Color Differences for Acceptability, which is of extended application in the textile industry [9].

A Color Eye 7000-A spectrophotometer (Gretag-Macbeth) with color integration sphere and colorimetry software was used. This equipment simulates the color appreciation by the human eye and translates it into a value or coefficient of color difference (ΔE) that, according to the method used, should not be greater than unity, that is, $\Delta E < 1.0$. This value less than 1.0 represents a color difference that a common observer could not perceive, while $\Delta E > 1.0$ is a color difference that any person can detect and, therefore, implies that the result of a dye is not acceptable [9].

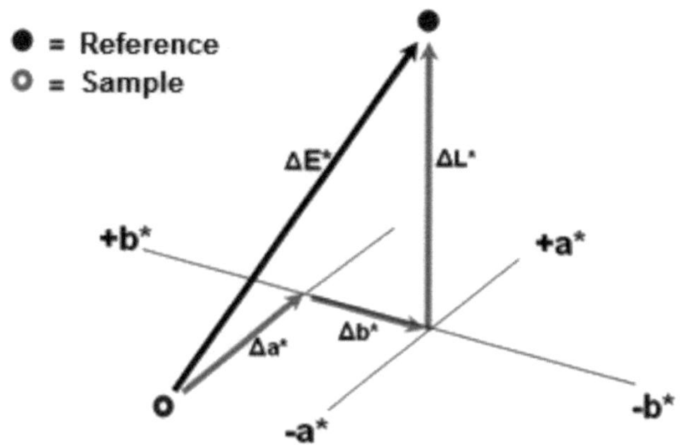

Figure 3.
Graphic representation of color difference, ΔE.

The principle on which the method of determination is based is of a psychophysiological nature and is known as Young's Theory. It refers to the fact that the human eye perceives color through three types of receptors and considers the sensory response as inscribed in a Cartesian space, where the axis L corresponds to the opposite pair black-white (−L corresponds to the color black and +L corresponds to the color white), the axis a corresponds to the pair green-red (−a corresponds to green and +a to red color), and the b axis corresponds to the yellow-blue pair (−b corresponds to blue and +b to yellow).

It is considered that each color tonality constitutes a point in the color space, and the color difference between two samples is calculated as the distance between the two points that represent the tonalities of the two samples compared, considering one of them as the standard. In this way, the color difference translates into a single numerical value that allows us to easily handle it. **Figure 3** illustrates a color difference between two samples (ΔE) within the color space.

3. Results

In all cases, the discoloration was carried out during the first hour, despite the low concentration of ozone used. The addition of inorganic auxiliaries accelerates the reaction of ozone with textile dyes. **Figure 4** presents the variation of the UV-Vis spectra for the RR141 during 60 min of ozonation. From the variation of the spectra of the RR141 dye during ozonation, it can be concluded that this compound is rapidly destroyed under the action of ozone. The tendency to discoloration during the ozonation process is very similar for both dyes. The studied compounds are rapidly destroyed under the action of ozone. **Figure 5** shows the discoloration of RB160 for three different concentrations at 615 nm.

The influence of the by-products of the degradation of the dyes on the dyes made with the treated water depends not only on the accumulation of these through the cycles of ozonation contamination but also of the dyeing class of the dye used in the dyeing with the treated water, as well as the chemical constitution of the latter. The results obtained in this study showed that for both dyes studied, analyzed at 50 mg/L, the color disappears after 30 min of ozonation.

Figure 6 shows the discoloration for the studied dyes at 50 ppm. Based on the preliminary analysis, it can be stated that the discoloration of the RR141 and RB160 is almost identical.

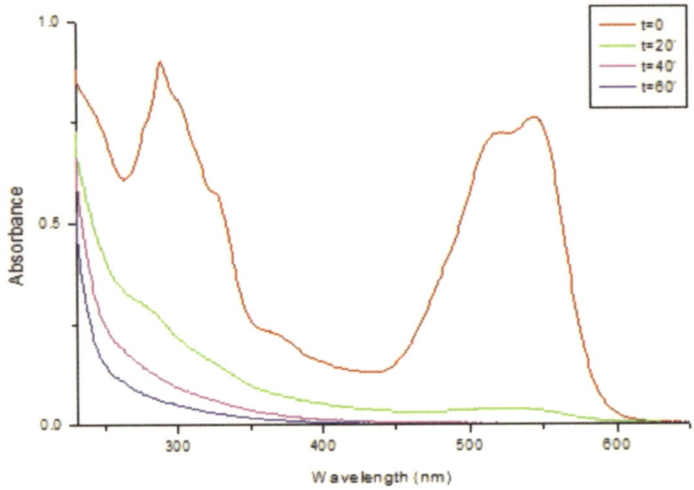

Figure 4.
Variation of the UV-Vis spectrum of RR141 during ozonation.

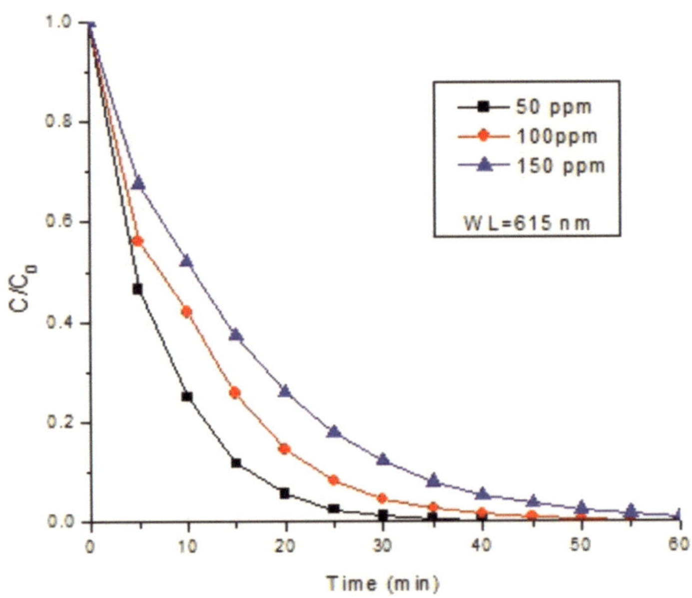

Figure 5.
Influence of concentration on discoloration of RB160 during ozonation.

RR141 has four amino groups and seven sulfonic groups, while RB160 has five amino groups and five sulfonic groups. Thus, taking into account the proportion between sulfonic groups and amino groups in the dye molecules, their reactivity and molecular weight, the observed discoloration rates are consistent. **Figure 7** shows the degradation of the dyes studied, during 60 min of ozonation and with an initial concentration of 50 mg/L.

The by-products of the two dyes require for their destruction of ozonation times greater than the 30 min necessary for the discoloration. Both curves are almost identical.

These results agree with the conclusions obtained by other researchers, who associate the presence of amino groups in the molecules of the dyes to a very fast reaction with ozone, while the presence of sulfonic groups increases the resistance

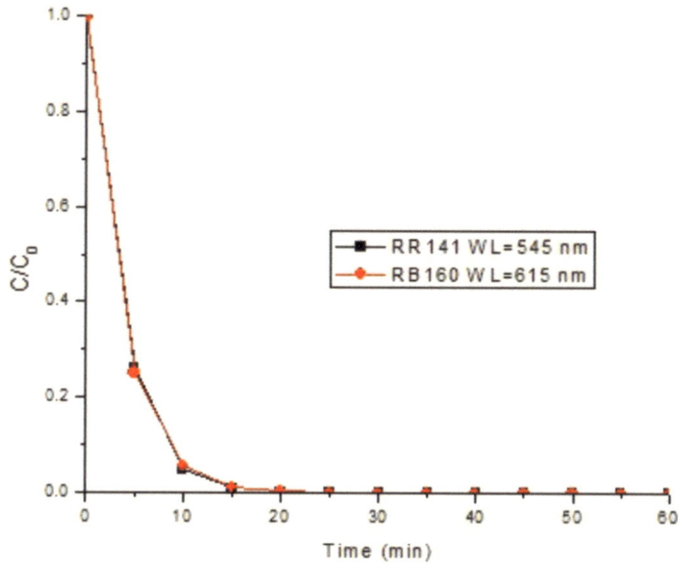

Figure 6.
Discoloration of the compounds studied at 50 mg/L.

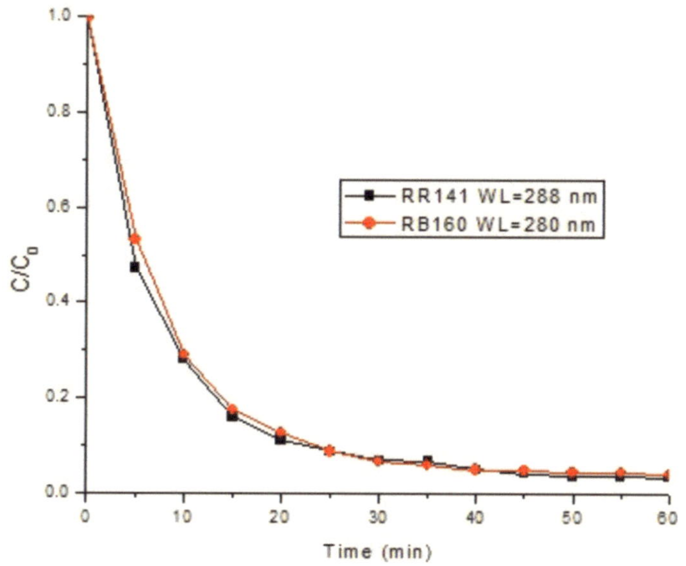

Figure 7.
Degradation of the dyes studied at 50 mg/L.

of the dye toward the attack with ozone [7, 11]. The composition of the intermediates and final ozonation products depends on the chemical structures of the dyes. Phenols, quinones, hydroquinones, and acids such as phthalic, muconic, fumaric, maleic, etc. have been identified among ozone intermediates. The final products that have been identified are sulfate and nitrate ions, elemental nitrogen, and oxalic and formic acids [11, 12].

If one takes as a starting point the mechanisms proposed in the literature for the degradation of organic compounds, the formation of maleic acid precedes the formation of oxalic acid and is carried out from the first minutes of ozonation. **Figure 8** shows the variation of the pH during ozonation of the dyes [7].

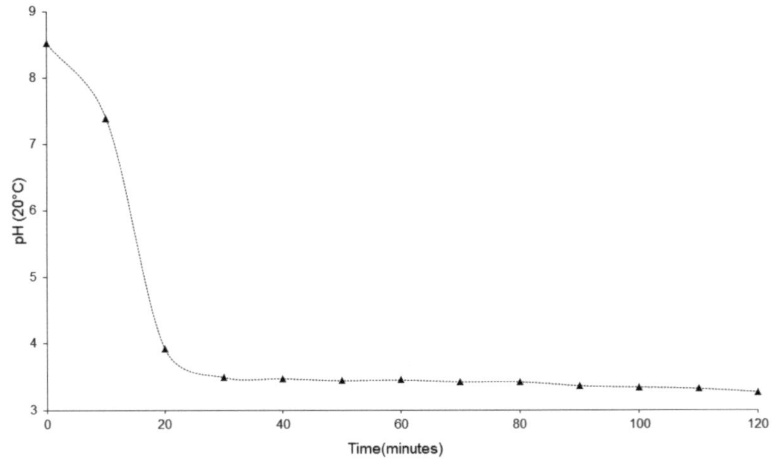

Figure 8.
pH variation during ozonation for RR141 at 50 mg/L.

The decrease of the pH in the ozonation of the dyes contributes to verify, in an indirect way, the formation of organic acids during the ozonation process. It is known that during the first minutes of ozonation, the formation of maleic acid is observed, the concentration of which then decreases markedly. However, the pH continues to decrease due to the subsequent formation of oxalic acid, which continues to form and accumulate throughout the ozonation process, although at a slower rate. It has been found that ozonation induces the denaturation and desulfurization of sulfo- and nitrophenols in the first stage of the reaction [13]. The increase in the conductivity of the dye solutions in ozonation serves as an indirect confirmation of the formation and accumulation of sulfate and nitrate ions, which in turn confirm the denitration and desulphurization of the dyes in the first step of the reaction. **Figure 9** shows the variation of conductivity for RR141 during ozonation [7].

During the process of dyeing textile materials with reactive dyes, the addition of large amounts of salt is necessary to promote the incorporation of the dye into the fiber. **Figure 10** shows the effect of salt addition on the ozonation of RR141 at 50 mg/L.

Once the dye is inside the fiber, it is necessary to add alkali, in order to promote the formation of the covalent bonds characteristic of this class of compounds. **Figure 11** shows the effect of adding sodium carbonate during ozonation for RR141 and **Figure 12** shows the effect of sodium hydroxide.

Due to this, the effluents of the dyeing with reactive dyes present considerable quantities of both salts and alkalis [7]. To study the influence of these inorganic substances, solutions of 50 mg/L of each of the dyes were prepared, and variable amounts of Na_2SO_4, Na_2CO_3, and NaOH were added to them before they were subjected to ozonation. The presence of inorganic auxiliaries in the dye solution has influence on the kinetics of discoloration. However, the concentration of these salts has no important influence on the kinetics. As an example, for RR141 and first-order reaction, rate constant is 0.0264 min^{-1} without addition of sodium sulfate, 0.0299 min^{-1} with addition of 10 g/L of salt, and 0.0303 min^{-1} with 20 g/L.

During the process of partial oxidation of the dyes with ozone, intermediate products are formed. These products accumulate through the recirculation cycles of the treated water and can influence the quality of the dyes made. **Figure 13** shows the accumulation of partial oxidation products through the consecutive recirculations for RR141.

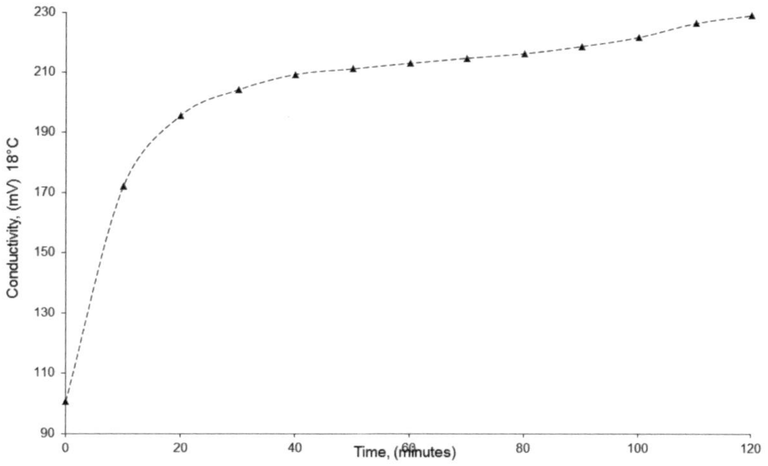

Figure 9.
Conductivity variation during ozonation for RR141 at 50 mg/L.

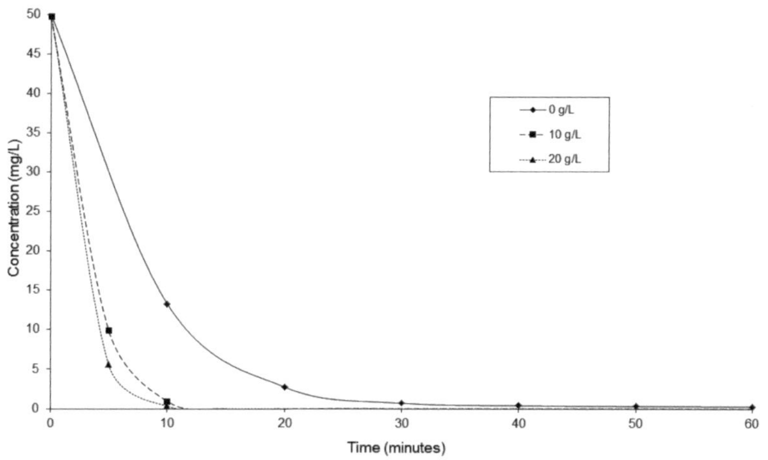

Figure 10.
Effect of salt addition during ozonation for RR141.

The water recycled up to five times by the procedure described in the methodology was used to make dyes on cotton cloth with direct dyes (AD80, AD50, and RD23) and with reactive dyes (AR84, RR141, and NR5). The determination of the quality of the coloration of these dyes was carried out by means of the AATCC method Test Method 173-2005 Calculation of Small Color Differences for Acceptability [9]. **Figure 14** illustrates the behavior of dyes made with direct dyes and reagents, in terms of the color difference coefficient (ΔE), when using RR141 solutions treated with ozone with up to five recirculation cycles of the pretreated water. A value of ΔE less than one means that the quality of the dye is acceptable.

Figure 14 shows that water contaminated with RR141 and pretreated can be used for dyeing processes with direct dyes during four cycles tested without significantly affecting the quality of the dye. The dye that best behaves in this case is Direct Blue 80 (AD80), which belongs to group B of the SDC classification, while Direct Yellow 50 (AD50) and Direct Red 23 (RD23), belonging to groups A and C, respectively, are very similar in their behavior. Regarding the reactive dyes, it is observed that it can be dyed with them without the coloration being affected for

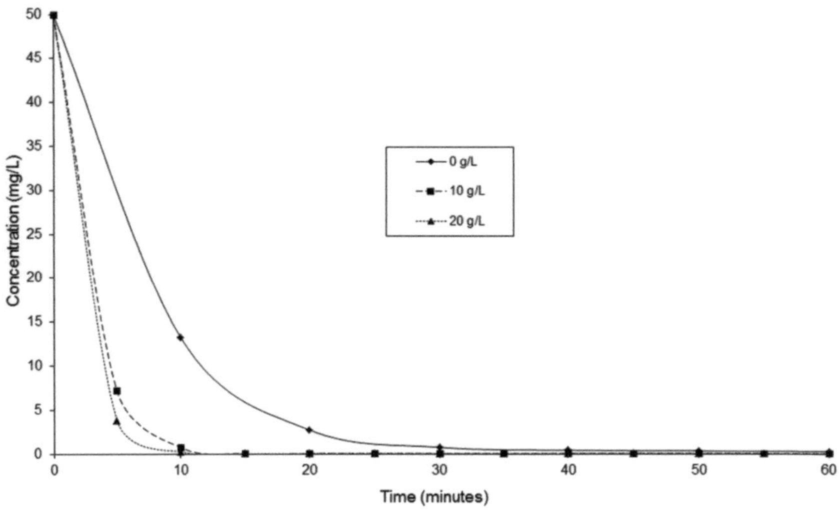

Figure 11.
Effect of sodium carbonate addition during ozonation for RR141.

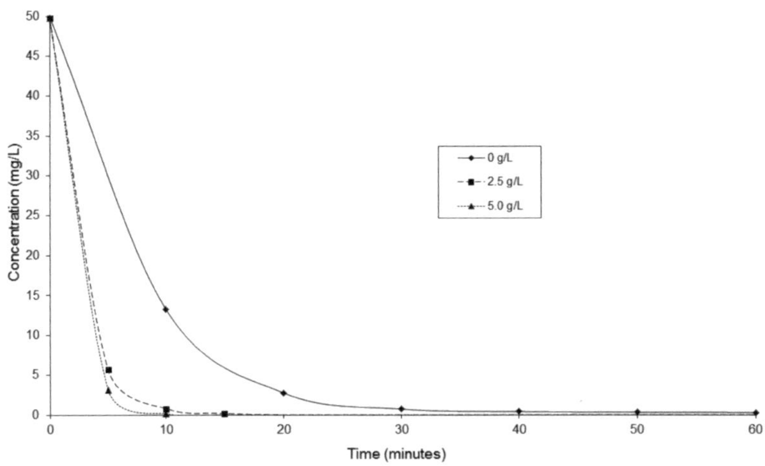

Figure 12.
Effect of sodium hydroxide addition during ozonation for RR141.

up to four cycles. Even though the NR5 presents differences of $\Delta E > 1.0$ from the second cycle, the results can be considered acceptable, since this dye is generally used at concentrations above 5% of the weight of the textile material, which makes it possible to hide the differences of coloration.

During the dyeing processes with reactive dyes, important quantities of sodium salts, mainly sodium sulfate, are added. The function of this salt is to neutralize the negative charge that the fiber acquires when impregnated with water. This function is performed by the sodium ion [11]. To verify that the sodium ion is not affected by the ozonation process, solutions were prepared as in the case of dye recirculations, with initial concentrations of 50 mg/L, both dye and sodium. The sodium ion content was analyzed by atomic absorption spectroscopy at 303.2 nm. **Figure 15** illustrates the result of the analysis for RR141 [14].

It can be observed that the content of sodium ion does not decrease during ozonation, which is the condition for the process to have an economic attraction

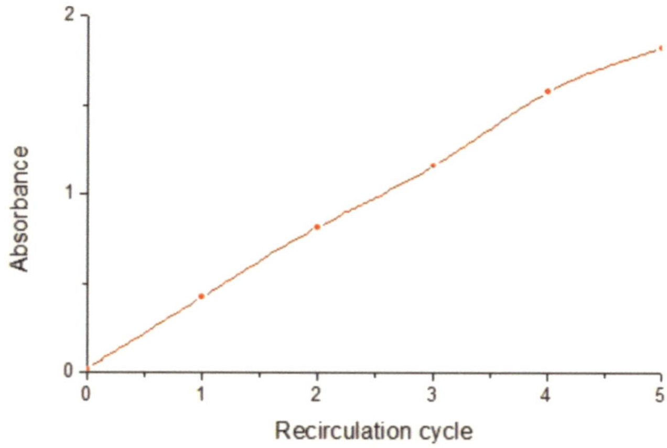

Figure 13.
Accumulation of products in the ozonation of RR141 with water recirculation.

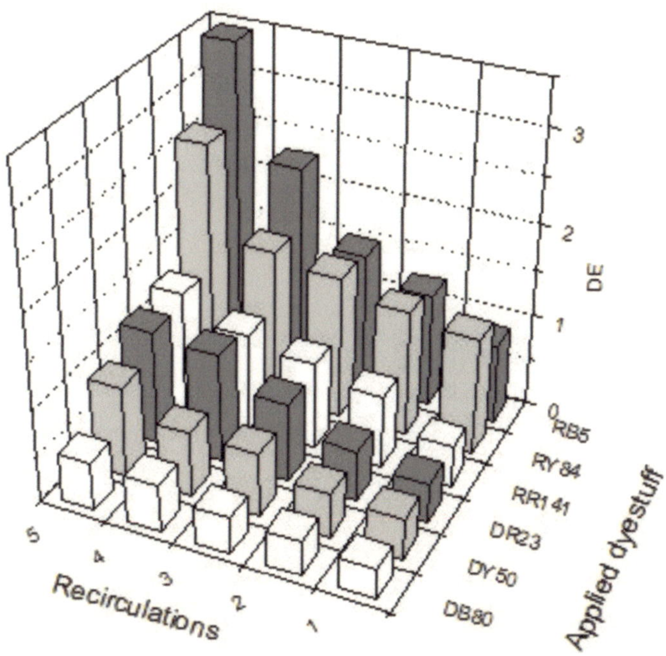

Figure 14.
Color differences obtained by dyeing cotton by reusing ozonated RR141 solutions.

due to the recovery of the salt contained in the exhausted dyeing bath. In fact, the sodium ion content increases slightly, due to the presence of ions of this type in the structure of the dyes and in the salts used in their manufacture to standardize them at commercial concentration.

A preliminary analysis by FTIR of the changes in the structures of the dyes during ozonation was carried out. In the area between 2000 and 4000 cm^{-1}, appears only one band at 3450 cm^{-1}; which corresponds to stretches of phenols or aromatic amines. **Figure 16** shows that region for RR141.

In the region between 600 and 2000 cm^{-1}, the dyes have several common absorption bands. One of them at 620 cm^{-1}, of bending of $SO_3^{=}$, together with another one at 1140 cm^{-1}, of stretching of bond S—O, confirms the desulfonation

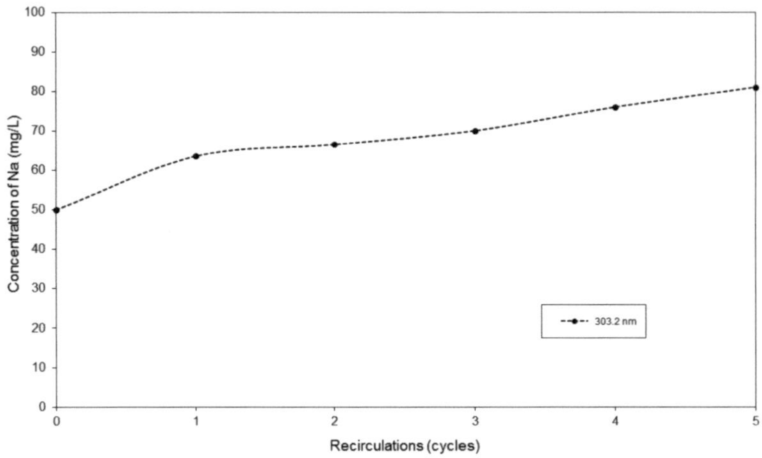

Figure 15.
Analysis by means of atomic absorption of the presence of sodium ion during the recirculation of water contaminated with RR141.

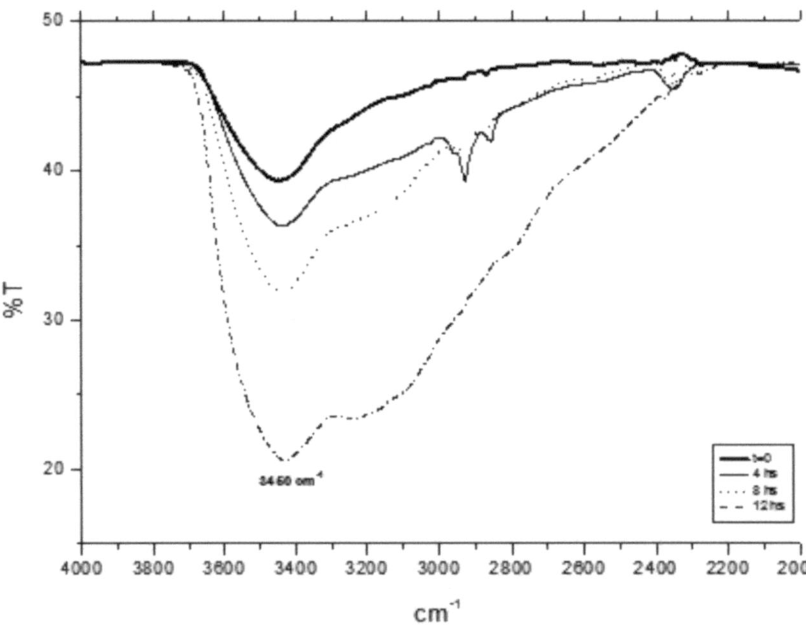

Figure 16.
Variation of the FTIR spectrum of RR141 during ozonation, between 2000 and 4000 cm⁻¹.

of the compounds during ozonation. Also, a band at 1050 cm^{-1}, of stretching of C—O bond, and a band located at 1400 cm^{-1}, of COO— ion stretching, confirm the formation of organic acids. Additionally, the dyes have absorption at 1500 cm^{-1}, corresponding to NO_2 formation, and at 1620 cm^{-1}, which indicates the formation of secondary amines.

RR141 show absorption at 760 and 1720 cm^{-1}. The first one corresponds to vibrations of the aromatic ring, while that of 1720 corresponds to the vibration of the -COOH group and indicates the formation of carboxylic acids. **Figure 17** shows region between 600 and 2000 cm^{-1} for RR141 [15].

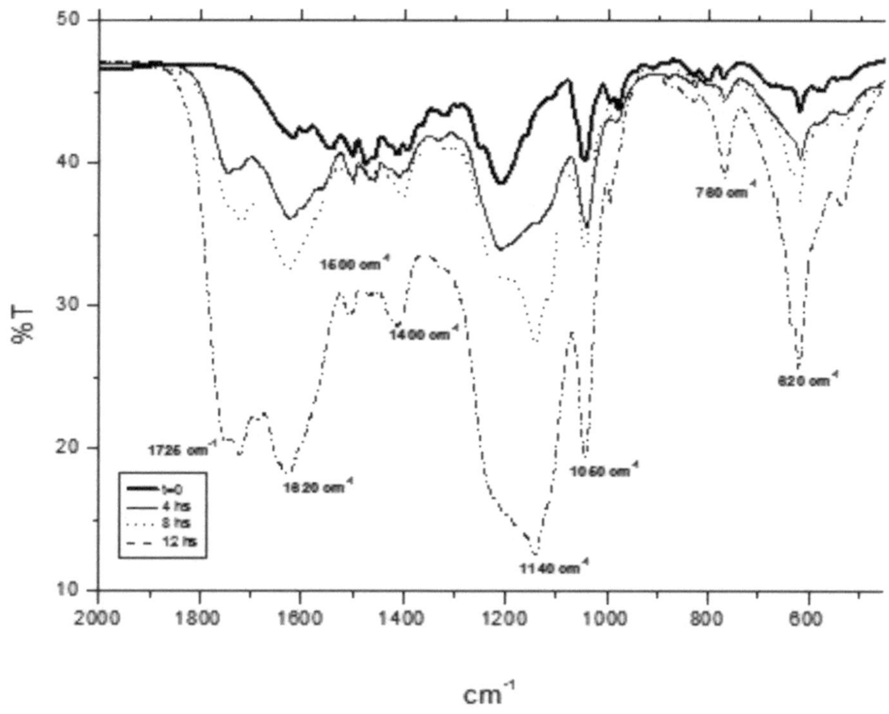

Figure 17.
Variation of the FTIR spectrum of RR141 during ozonation, between 600 and 2000 cm^{-1}.

4. Conclusions

According to the results, we can conclude the following:

1. The discoloration of the model solutions of RB160 and RR141 dyes by ozonation is carried out during the first 30 min of treatment.

2. The rate of the ozonation reaction for the dyes is different and depends mainly on the chemical structure of the dye used, the initial concentration of this, and other factors, such as the dyeing auxiliaries.

3. In addition to the discoloration, it is possible to decompose the intermediates present in the water, until obtaining biodegradable and nontoxic compounds such as organic acids, if the ozonation time is prolonged.

4. During the cycles of recirculation of the water, an accumulation of ozonation products is observed. However, these products allow water to be reused for several cycles, depending on the type of dyes used in dyeing with treated water. In addition, such ozonation products can be degraded by longer ozonation times (>120 min).

5. A water recirculation scheme was proposed in dyeing processes of textile materials, both with reactive dyes and with direct dyes, which does not affect the quality of the dyeing result.

6. It was shown that the proposed recirculation scheme works properly using water that has been premixed for a period of 10 min, up to four cycles, for the case of dyeing with direct dyes, and up to four cycles, for the case of dyeing with reactive dyes.

7. Likewise, it was demonstrated that the sodium ion content is not affected by ozonation during the water recirculation cycles. In fact, it increases slightly, due to the sodium ions contained in each dye, both in its molecular structures and in the salts used to standardize them.

8. Some of the advantages of the reuse of water treated by ozonation according to the proposed scheme are the following:

    - Elimination of a good part of the color in the discharges into the environment, decreasing visual pollution.
    - Elimination of the use of other treatments, such as the addition of polymers, or the use of membranes that are effective, but at the same time costly, and do not eliminate the problem.
    - Improvements in the final treatment of wastewater, because the decomposition products of the dyes are more biodegradable than the dyes themselves.
    - Reduction of the consumption of water and salts used in the dyeing, as well as reduction of discharges of salts to the environment. These salts are sulfate and sodium chloride, mainly, and are used during the dyeing to facilitate the diffusion of the dye in the textile material.
    - The recovery of the salts and water used, by itself, makes economically attractive the reuse of spent dyeing baths, especially in the case of reactive dyes, where the salt concentration in the dyeing bath becomes approximately 10% w/v.

Acknowledgements

P. Colindres-Bonilla thanks the economic support of National Institute of Technology of Mexico (Project: 6476.18-P).

Conflict of interest

The author declares that he has not conflict of interest.

Author details

Pablo Colindres Bonilla
National Institute of Technology of Mexico, Mexico City, Mexico

*Address all correspondence to: pcolindres@ipn.mx

References

[1] Aspland JR. Textile Dyeing and Coloration. USA: American Association of Textile Chemists and Colorists; 1999

[2] Shore J. Colorants and Auxiliaries, Colorants. Vol. 1. England: Society of Dyers and Colourists; 2002

[3] Broadbent AD. Basic Principles of Textile Coloration. England: Society of Dyers and Colourists; 2001

[4] Hunger K. Industrial Dyes, Chemistry, Properties, Application. Germany: Wiley-VCH; 2003

[5] Christie RM. Environmental Aspects of Textile Dyeing. England: Woodhead Publishing in Textiles; 2009

[6] U.S. Environmental Protection Agency. Best Management Practices for Pollution Prevention in the Textile Industry. USA: EPA; 1996

[7] Razumovskii SD, Zaikov GE. Ozone and its Reactions with Organic Compounds. The Netherlands: Elsevier; 1984

[8] Society of Dyers and Colourists and American Association of Textile Chemists and Colorists. Colour Index. 3rd ed. England: S.D.C. and A.A.T.C.C.; 1971

[9] American Association of Textile Chemists and Colourists. A.A.T.C.C. Technical Manual 2007. USA: A.A.T.C.C.; 2006. 82 p

[10] Gomes AC, Nunes JC, Simões RMS. Determination of fast ozone oxidation rate for textile dyes by using a continuous quench-flow system. Journal of Hazardous Materials. 2010;**178**:57

[11] Perkins WS. Oxidative decolorization of dyes in aqueous medium. Textile Chemist and Colorist. 1999;**32**:33

[12] Faria PCC, Órfão JJM, Pereira MFR. Activated carbon and ceria catalysts applied to the catalytic ozonation of dyes and textile effluents. Applied Catalysis B: Environmental. 2009;**88**:341

[13] Zhang J, Lee KH, Cui L, Jeong TS. Degradation of methylene blue in aqueous solution by ozone-based processes. Journal of Industrial and Engineering Chemistry. 2009;**15**:185

[14] Perkin-Elmer. Analytical Methods for Atomic Absorption Spectroscopy. 1st ed. USA: Perkin-Elmer; 1996

[15] Silverstein RM. Spectrometric Identification of Organic Compounds. 7th ed. USA: John Wiley & Sons Inc.; 2005

Chapter 4

Utilization of Cotton Spinning Mill Wastes in Yarn Production

Tuba Bedez Ute, Pinar Celik and Memik Bunyamin Uzumcu

Abstract

Efficient use of natural resources and utilization of recoverable wastes are getting more and more important day by day since recovering wastes have both economic and environmental benefits. As the source material costs constitute the majority of the yarn production costs, decreasing raw material costs provide considerable advantages for spinners. From the point of textile manufacturing, various production wastes can be reused in textile industry. In each step, from ginning (for cotton fibers) to end product formation, recyclable/recoverable waste materials are generated. However, mainly polyester products are recycled (r-PET) and used again in textile industry by 100% or in blends with other man-made or natural fibers. Compared to research on r-PET, recovered cotton fibers inspired interest recently. The main objective of this study is to fill the gap in the literature via investigating the properties of the yarns produced with recovered cotton wastes, generated in different sources. For this purpose, spinning mill waste types were selected. In this experimental study, different waste types (card waste, blowroom waste, and fabric waste) and blending ratios were used. As a conclusion, the effect of waste type and blend ratio on the physical and mechanical properties of the yarns and the fabrics, produced with virgin and waste cotton fibers, were analyzed.

Keywords: waste, recovering, recycling, spinning mill, fiber, cotton

1. Introduction

Every being in this world has an expiration date, even the world itself possibly has one. This situation is the same for man-made products too. We produce them, use them, and try to find a way to get rid of them, when the time comes. One of the hardest questions of today comes to mind at this point: How will we manage the resultant waste of the products we created? Should we dump the waste to proper waste yards and reuse or recycle them? The answers to these questions are crucial. Scientists, governments, and local authorities work for finding answers to these questions. Wastes can be a problem for local authorities because of their environmental effects.

There are different categories of wastes. According to Australian Waste Report 2016, these categories are masonry materials (asphalt, bricks, rubble, etc.), metals (steel, aluminum, etc.), organics (food, garden organics, etc.), paper and cardboard (liquid paperboard, magazines, etc.), plastics (PET, HDPE, PP, etc.), glass, hazardous wastes, fly ash, and other wastes (including textiles and leather) [1]. Some of these waste types can be recycled and utilized as raw materials in same type of products they belong in or in different products.

Textile wastes can be divided into two main groups: production wastes and postproduction wastes. Production wastes are basically raw materials of each production step which cannot be put into end product due to different reasons. For yarn spinners, these wastes can occur during cleaning of the fibers or combing out short-staple fibers from the long ones in combing machine, etc. These clean/unclean wastes in fiber form or not can be reused. After spinning mill, there are wastes in yarn and fabric forms, and they need recycling to be put again in production. Postproduction wastes are generally worn out cloths, which can be recycled and may be used again in textiles or utilized in other products.

Textiles include different raw material (fiber) types. Fibers used in textiles are categorized into two main groups, which are natural and man-made fibers. Most known examples for natural fibers are cotton (seed fibers), wool, silk (animal fibers), flax (bast fibers), sisal (leaf fibers), and asbestos (mineral fibers). On the other hand, polyester, nylon, acrylic (which are synthetic fibers), modal, viscose rayon, and acetate rayon (which are regenerated fibers) are some of the examples for man-made fibers [2]. Thereby, textile wastes have a great variety of raw material sources. These wastes can be recycled or reused in different products. In 2017, global fiber production exceeded 100 million mt. Polyester has around 51% of total global fiber production. The second most important fiber is cotton, and it has approximately 25% of total global fiber production [3].

Textile wastes can be recycled/reused in textiles or other products. Other product wastes can also be utilized in textile production. One of the most known examples for this is PET bottles. PET bottles are collected, are recycled, and can be used in textile products as "r-PET fiber." r-PET fibers can be used in yarn production, as 100% or in blends, thereby in most of the textile structures. There are various studies about this topic. These studies cover spinning of the fiber, properties of the fiber [4], properties of yarn, and fabric produced from this fiber and all [5]. Some of the researches are focused on using textile wastes in different products. Mishra et al. used textile wastes to produce composites and tested the properties of these composites [6]. Briga-Sa, Binici, and El Wazna used textile wastes as insulation materials, and Briga-Sa indicated that they got results similar to polystyrene (XPS) and mineral wool [7–9]. Shukla used PET fiber wastes to synthesize new chemicals which can be used in different fields [10]. These examples show that textiles are generally sustainable materials. There are too many studies dedicated on this topic.

Liquid waste and solid waste are generated during the production of textiles. Especially agriculture of cotton fiber, which is the subject of this study, and the evaluation of the solid wastes that occur during yarn spinning are important for sustainability and environment. These topics are really important for the future, considering the increasing world population and decreasing agricultural areas. Moreover, large amounts of water are consumed; pesticides and synthetic fertilizers are used during cotton growing. Especially pesticides have negative effects on human health. The recycling/reusing of wastes occurring at every stage of textile production will be positive in terms of reducing the environmental effects putting them again in the production chain. Consumers are also becoming more conscious about these effects and they seek environmentally friendly "green products" [11]. This should be considered by the producers.

Due to increasing fiber production, the amount of pre-consumer and postconsumer textile wastes is increasing. According to a study Pinheiro and de Francisco conducted with Brazilian clothing manufacturers, 167,850 kg of cotton was consumed with these manufacturers, and 19,086 kg of cotton waste occurred in their productions. This means 11% of the raw material was left out as waste [12]. Moreover, as the amount of textile solid waste increases, the evaluation of these wastes becomes more important. Therefore, the terms of sustainability and circular economy issues come to the forefront in textile industry. This chapter is focused on spinning mill wastes of textile wastes. Information about how textile wastes occur in spinning mills and utilization of them in textiles are given. For this reason, informations about spinning mill and wastes occurring in spinning mill were given, initially.

1.1 Short-staple yarn spinning

Spinning is defined by Barker as the "*art of throwing a number of more or less short fibres together in such a way that, being drawn out to form a comparatively fine filament*" [13]. In this process, one of the main defining parameter is fiber length. According to fiber length, machinery and their adjustments that should be used are determined. In textile yarn manufacturing, two main systems are used depending on fiber length: short-staple and long-staple spinning systems. In principle, fibers up to 60 mm in length are spun in short-staple spinning systems, and fibers with lengths over 60 mm are spun in long-staple systems. Short-staple fibers are generally processed dry using mechanical means, and the spinning systems used for this types of fibers are also known as cotton spinning system [14].

From the field, seed cotton moves to gins for separation of lint and seed. This is the first step in which cotton wastes occurred. Cotton gin wastes consist of sticks, leaves, burs, soil particles, mote, cotton lint, and other plant materials [15]. These wastes can be used in different areas such as chemical industry (e.g., soaps), livestock industry (e.g., animal feed), or food industry (e.g., cotton seed oil).

After harvest and ginning, cotton fibers are compressed and bales are formed. For this reason, the first step in a cotton spinning mill is opening. This process is needed in order to clean effectively and form slivers in which individual fibers are oriented very closely to sliver axis [16]. Most of the opening and cleaning is carried out in blowroom machinery. However, card has an important role in opening and cleaning. Most of spinning mill wastes occur in these machinery.

Parallelization is carried out subsequent to opening and cleaning processes. It is really important to force fibers to place as parallel to each other as possible in sliver to spin a good quality yarn. Machine mainly responsible for this process is draw frame which also takes care of attenuation and doubling of slivers. Fibrous waste amount of these machines is lower compared to the rest of the spinning mill machinery.

To produce some cotton end products especially in which fine yarns are used, yarns with better properties are needed. One of the ways to do so is to remove some fibers that are much shorter than the mean of the distribution from slivers [16]. This process is carried out with combing machine. In **Figure 1**, spinning machinery line with combing machine is given. Cotton fibers longer than 27 mm are generally used after combing up to 13% which is sufficient for good quality yarns [16]. This means waste ratio of this process is high.

During roving and spinning (for ring spinning), some fibers cannot enter yarn or roving body, and fiber fly is formed. These fibers are sucked by pneumatic systems that are placed after delivery rollers, and they are collected in the machine. Moreover, if end breaks happen in yarn or roving, the same system collects flowing fibers after leaving delivery rollers till operator's intervention. In open-end rotor spinning systems wastes can occur in opening rollers which also are responsible for cleaning.

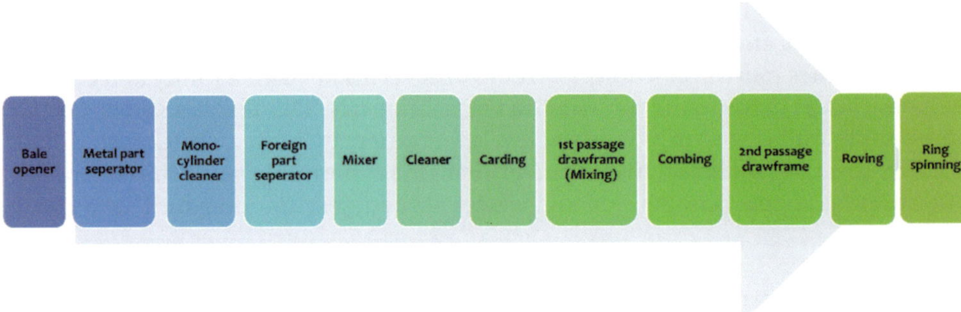

Figure 1.
Opening and cleaning lines of combed ring spinning [17, 18].

1.2 Cotton wastes in spinning mill

Klein classified cotton fibers used in short-staple spinning as virgin fiber (from ginning mill), clean waste, comber waste, recycled fibers from dirty waste, and fibers torn out of hard waste (roving, yarn, and twisted threads) [19]. Spinning wastes and their sources are given in **Figure 2**.

Broken ends of sliver, lap, web, and filter strippings from draw frame, roving frame, ring spinning frame, and rotor spinning frame are known as clean waste, having more than 95% of good fiber. Comber and roving wastes' good fiber ratio is around 95–97%. Wastes generated in blowroom machines, and cards are dirty wastes with 35–55% good fiber ratio. Besides, another dirty waste, flat and filter stripping waste, has a higher amount of good fiber (65–80%). As the waste fibers are processed in different number of machines and therefore stressed fibers, their good fiber content is less than virgin fibers. For this reason, spinners prefer to feed the waste fibers into normal spinning process, in a controlled manner, with a constant percentage in order to avoid quality variations. Generally, wastes arising in the mills can be returned to the same blend from which they arose; comber wastes are mostly used in rotor spinning. In carded ring-spun yarn and fine rotor-spun yarn production, waste fibers can be used, up to 5%, but for combed yarns, waste fiber ratio can be lower, up to 2.5%. Higher waste fiber amounts can be used for medium and coarser rotor yarns, about 10 and 20%, respectively.

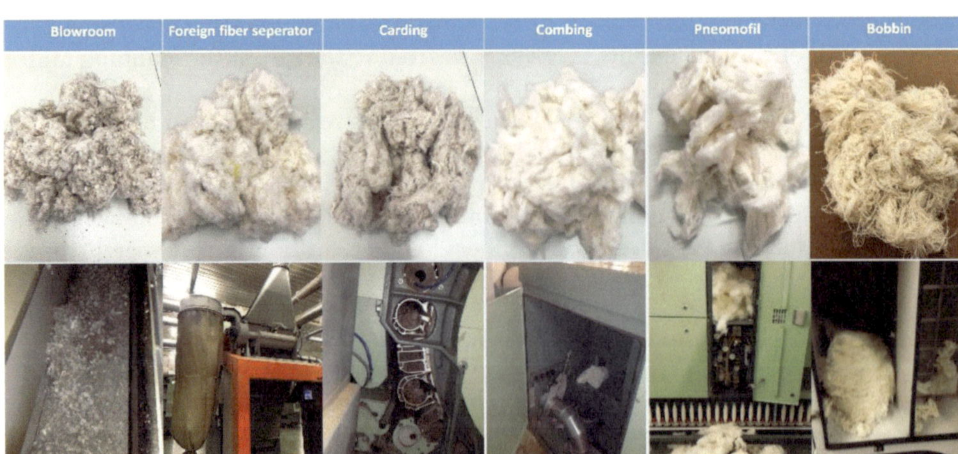

Figure 2.
Spinning mill wastes in different sources.

According to Klein, waste generation in spinning mills of industrialized countries differs from machine to machine. The waste mostly occurs in comb machine, as one of its primary functions is to remove short fibers which are called comber wastes. Lowest waste creating machine in the spinning mill is draw frame. If the spinner is producing carded yarns, blowroom is responsible for the most waste generation. However, in blowroom machinery and card, shorter fibers have the highest ratio in the resultant waste despite ring spinning frame. In ring spinning frame and comber, waste consist of longer fibers (up to 1 1/2″) in comparison with card and blowroom machinery [20].

As the raw material costs constitute the majority of yarn production costs, spinners prefer to use waste fibers in the blends. In addition to saving raw material costs and the requirement for efficient use of limited raw material resources, the possibility of using higher degree of cleaning in the blowroom machines is the other advantage. On average, about 6–8% primary waste which is composed of 50% good fiber can be expected. About 90% of the good fiber elimination can be recovered as secondary raw material that is still containing 6% trash.

Many researches focused on cotton spinning wastes. Wulfhorst concluded that these secondary raw materials can be blended with raw materials with a percentage up to 20%, without noticeable changes in yarn quality [21].

Duru and Babaarslan have determined an optimum opening roller speed for spinning different types of 60/40 polyester/waste blended rotor yarns. Waste fibers were obtained from cotton noil, recycled fibers, flat waste, etc. and were blended and processed on the traditional short-staple (carding) system. They have found that yarn strength, unevenness, and hairiness values decreased with the increase of the opening roller speed, and the optimum opening roller speed was 7000 rpm for both yarn properties and cleaning effects [22].

The effect of some rotor spinning parameters, such as rotor diameter, rotor speed, separator angle, navel type, opening roller speed, and yarn linear density on the level of yarn hairiness of the rotor-spun yarn produced from ginning wastes in different waste proportions, was investigated by Hasani et al. by the Taguchi method. They have found that, for 35/65 cotton waste-spun yarns, rotor diameter was the most effective factor, followed by yarn linear density, separator angle, opener speed, navel type, and rotor speed factors, respectively. Yarn linear density was the strongest effect for 50/50 cotton waste-spun yarns, and it was followed by rotor diameter, rotor speed, opener speed, navel type, and separator angle factors. Rotor speed and opener speed were found the least effective factors of all, for 65/35 cotton waste-spun yarns [23]. According to their following research, the rotor diameter, yarn linear density, and the navel type had the largest, and the opening roller speed had the lowest effect on yarn hairiness, for all waste ratios. Furthermore, they have found that yarns produced with higher waste proportions had higher hairiness values [24].

Khan and Rahman have focused on the effects of rotor spinning parameters too. They have studied the effects of rotor speed, combing-roll speed, and type of recycled waste on rotor yarn quality and end breakage, using response surface methodology. They have collected spinning wastes from different positions of ring spinning process (flat strips, noil, filter waste, and Pneumafil) and used after recycling, except Pneumafil. They have reported that yarn strength, elongation, imperfections, and end breakage rate can be improved by using Pneumafil waste, mostly in the range of 5–25%. The negative impacts of rotor speed on yarn imperfections and end breakage can be minimized also. In their study, rotor and combing-roll speed of 85,000 rpm gave better results in terms of yarn strength and elongation; on the other hand, the lowest end breakage rates were found at lower speeds [25].

Khan et al. have also focused on predicting the unevenness, imperfections, strength, and elongation properties of cotton/waste blended rotor yarn, using

Taguchi OA experimental design. They have used blend ratios, blending technique, cylinder speed, and rotor speed as predictors. They have collected flat strips from carding machines and recycled them. Besides, comber noils, untwisted roving wastes, and Pneumafil wastes were directly used. They have produced virgin cotton/spinning waste blended yarns with two different levels of blend proportion (17/83 and 33/67) for both blowroom blending and draw frame blending. Their model showed that blend ratio and rotor speed are the most influencing factors for yarn quality. Reducing rotor speed improved the yarn evenness, imperfections, and strength. Yarn strength was found lower with draw frame blending, but on the other hand with this blending type, high portion waste containing yarns' evenness, imperfection, and elongation values are better than blowroom blending [26].

Halimi et al. investigated the effect of waste ratio (0, 12.5, 25, 37.5, 50, 62.5, 75, and 100%) and spinning parameters (rotor type, rotor speed, and opening roller speed) on the rotor yarn quality. Cotton wastes were collected from opening-cleaning machines and cards processed to reduce impurities. Greek cotton was chosen due to the suitable fiber length for blending with recovered fibers. They have reported that yarn appearance, level of irregularity, and the yarn uniformity did not affect up to 25% waste ratio [17, 18].

Celep et al. investigated the thermal comfort properties of the single jersey fabrics produced from virgin/recycled cotton fiber-blended open-end rotor yarns (100/0, 50/50, and 0/100). Recycled cotton fibers were obtained from fabric scraps from garment industry. They have found that fabrics containing recycled cotton fibers show higher thermal resistance and lower thermal conductivity, thermal absorptivity, and air permeability and give a warmer feeling at first touch [27].

Recycled denim fabrics by using recycled cotton fibers (varying from 30 to 85% blend ratio), recycled PES fibers, Tencel fibers, and virgin cotton fibers were produced. They have found that recycled fiber-blended yarns have higher unevenness, IPI faults, and hairiness comparing to the standard yarns, resulting with noticeable nubs in fabric surface. This case did not affect fabrics' physical and mechanical properties significantly but provides better abrasion resistance contrary to expectations from recycled products [28].

Yılmaz et al. have focused on the effects of different waste cotton fiber types and the amount of waste in the blends (varying from 5 to 40%), on the quality of conventional ring and OE-rotor yarns. They have used preparation process wastes (blowroom and carding) and Pneumafil wastes (sucked on the draw frame, roving frame, and conventional ring spinning machines). They have found both for conventional ring-spun and OE-rotor yarns that the blends containing pneumafil wastes resulted with better yarn properties, while the blends with blowroom and flat wastes caused worse yarn qualities. In general, flat waste fiber blends have higher neps and hairiness values. They have concluded that when the waste percentage was increased from 5 to 40%, yarn irregularity values increased up to 37 and 16%, yarn hairiness increased by about 21–22%, yarn tenacity values decreased by 22, and 52%, breaking elongation decreased by 7 and 38%, for ring-spun and rotor yarns, respectively. As expected, with the usage of waste fiber, the most deteriorated yarn properties were yarn unevenness in conventional ring-spun yarns and tensile properties in rotor yarns. Their findings showed that the effect of waste fiber usage on knitted fabrics' pilling behavior was significant and Pneumafil fiber blends increased the pilling resistance, while for other waste fiber blends, on the contrary [29].

Béchir et al. evaluated the effect of recycled fiber ratio and number of recycling passages on the yarn quality and predicted the quality of the blend using a mathematical approach. They have concluded that recycling process of cotton waste with four passages gave an optimal global quality of fibers. Unevenness and IPI values of blended yarns increased with the increasing of recycled fiber ratio in the yarn [30].

Demiroz Gun et al. studied the dimensional and physical properties of the socks made from the blend of reclaimed fiber and polyester fiber. As a result of the study, they reported that reclaimed fiber can be used for production of socks, when the blend of virgin polyester fiber and reclaimed fiber are used, and thus acceptable quality can be obtained [31].

Yuksekkaya et al. used "yarn quality index" value, which was defined by Yunus and Rahman (as Yarn Quality Index = (Strength × Elongation)/ Evenness), for evaluating yarn quality in their study [32, 33]. It can be said that the most important parameter in order to evaluate yarn spinnability is the yarn tensile strength. They reported that generally, yarns produced from recycled fibers displayed better properties in terms of unevenness, yarn imperfections, and yarn quality index value. On the other hand, they found that yarn tensile strength and fabric burst strength were lower for recycled yarns and fabrics when compared to virgin ones [32].

2. Material and methods

The main objective of this study is filling the gap in the literature via investigating the properties of the yarns produced with cotton wastes, generated in different sources. For this purpose, different waste types (card waste, blowroom waste, sliver waste, and recycled cotton fiber from ecru fabric waste) and waste ratios (10/90, 30/70, and 50/50) were used in rotor spinning. As a pre-treatment and recycling process, all waste types were processes in Micro Dust and Trash Analyzer (MDTA) in two passages. Simultaneously, trash and fiber content of these wastes were analyzed, and the fiber length specifications of wastes after recycling process were evaluated by length control tester (Textechno). Length control tester is a mobile device for use in the spinning mill, developed for the measurement of fiber length parameters on slivers or cotton in tuft form. The test results of length control tester give information about the mechanical stresses which the fibers undergo in the manufacturing process and optimum settings of card, draw frame, or combing machines [34]. Besides, fiber contents of the wastes were analyzed with MDTA, and test results are given in **Table 1**. Trash analysis of the virgin cotton was performed from sliver form, resulting with high ratio of fiber content. Subsequently, wastes were blended with virgin cotton fibers (in sliver form) with the same machine. Then, Ne 20/1 open-end yarns were spun by using these slivers on Rieter open-end machine (R40). Single jersey fabrics were knitted by using Mesdan Lab Knitter with the same tightness factors under constant machine settings.

Waste type	Fiber content (%)	2,5% SL* (HVI) (upper length) (mm)	50% SL (mm)	ML* (AFIS) (LCT length) (mm)	Fiber hooks (%)	SFL* (mm)	SFA* (%)	SFC* (AFIS short fiber amount) (%)	Staple gradient (HVI uniformity ratio)
Blowroom	45.74	25.27	9.24	15.62	5.3	7.26	56.4	38.0	38.4
Card	82.44	21.69	6.71	7.79	1.2	2.81	78.0	76.6	33.0
Sliver	82.11	27.63	11.62	11.04	0.4	11.04	44.5	15.0	42.0
Fabric	80.89	22.64	8.46	13.36	8.4	6.07	57.2	47.6	39.6
Virgin cotton	99.41	27.67	12.50	20.59	4.8	10.38	42.5	17.4	42.5

**is corresponding to the test result given in brackets.*

Table 1.
Fiber specifications.

The standards of yarn and fabric tests applied were given below:

- Bursting strength (EC 37 hydraulic bursting strength tester) TS 393 EN ISO 13938-1
- Air permeability (Textest FX 3300) TS 391 EN-ISO 9237, ASTM D737
- Pilling test (ICI Pilling Box-7000 tours) TS EN ISO 12945-1
- Fabric thickness (SDL Digital Thickness Gauge) TS 7128 EN ISO 5084
- Yarn strength test (Lloyd yarn strength tester) TS 245 EN ISO 262, ASTM D 2256
- Yarn evenness test (Uster Tester 5) ISO 16549

The effect of waste type and blend ratio on the physical and mechanical properties of the yarns and the fabrics, produced with virgin/waste cotton fibers, were statistically analyzed by using SPSS.

3. Results and discussions

According to the results given in **Table 1**, the sliver waste and virgin cotton fibers have similar fiber specifications, because of being a clean waste as mentioned previously.

Physical and mechanical properties of the yarns were tested, and the test results were given in **Table 2**. In addition, the fabric thickness, bursting strength, air permeability, and pilling values of the fabrics were tested, and the test results were

Waste type	Waste/virgin cotton	Uster (CV%)	Thin places (−50%)	Thick places (+50%)	Nep values (+280%)	Yarn hairiness (H)
Blowroom	10/90	16.88	9.17	106.67	105.00	5.53
	30/70	17.95	361.67	245.83	412.50	5.76
	50/50	19.47	18.33	385.00	383.33	6.15
Card	10/90	16.53	5.83	50.00	23.33	5.84
	30/70	16.64	10.00	75.00	59.17	6.37
	50/50	15.48	4.17	78.33	62.50	6.82
Sliver	10/90	15.81	16.67	22.50	8.33	5.24
	30/70	15.68	28.33	27.50	11.67	5.14
	50/50	16.94	29.17	26.67	6.67	5.05
Fabric	10/90	16.30	5.00	75.83	40.00	5.58
	30/70	16.27	1828.33	186.67	1077.50	5.86
	50/50	16.26	52.50	209.17	125.00	6.04
Virgin cotton	0/100	15.26	8.00	26.00	6.00	5.32

Table 2.
Yarn test results.

Waste type	Waste/virgin cotton	Fabric thickness (mm)	Air permeability (l/m²/s)	*Pilling (1–5)
Blowroom	10/90	0.78	1440	2–3
	30/70	0.75	1274.8	3–4
	50/50	0.78	1282	3
Card	10/90	0.78	1444	5
	30/70	0.76	1412	5
	50/50	0.79	1256	5
Sliver	10/90	0.76	1758	4
	30/70	0.75	1830	4–5
	50/50	0.73	2052	4
Fabric	10/90	0.76	1552	4
	30/70	0.82	1410	3–4
	50/50	0.78	1534	3
Virgin cotton	0/100	0.67	2786	4–5

**1, worst–5, best.*

Table 3.
Fabric test results.

given in **Table 3**. For better understanding and evaluation, yarn tenacity and bursting strength values of different samples are given in **Figure 3**, respectively.

The effects of waste amount and waste type on yarn and fabric properties were statistically analyzed (**Table 4**). The effects of waste amount on yarn and fabric properties were found statistically significant, except yarn unevenness values. According to the Student-Newman-Keuls test, the number of thick places and nep values of the yarns produced with blowroom wastes were statistically significantly highest, of all. In terms of yarn hairiness, the highest values belong to the yarns produced with card waste, for all blend ratios. Card waste has the highest short fiber amount, resulting with the yarns' hairiness. For higher waste ratios (50/50), clean sliver waste containing yarns show better yarn tenacity, close to 100% virgin cotton.

One of the main factors affecting measured fabric properties is yarn hairiness. Yarn hairiness increases, and bursting strength decrease with the increase of waste ratio, except when the waste type is sliver. Sliver waste/virgin cotton-blended yarns have the lowest yarn hairiness values and highest bursting strength, even better than virgin cotton, because of better fiber specifications.

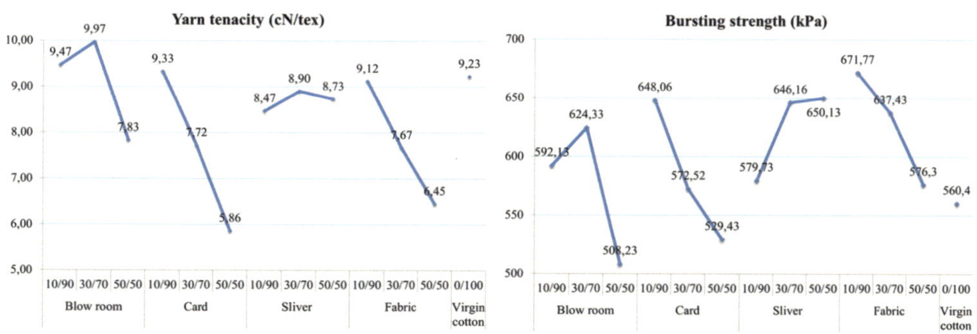

Figure 3.
Yarn tenacity and bursting strength values of different samples.

	Significance (p)			Significance (p)			
	Waste/virgin cotton ratio			Cotton waste type			
	10/90	30/70	50/50	Blowroom	Card	Sliver	Fabric
Yarn evenness (CV%)	0.541	0.541	0.541	0.160	0.478	0.806	0.998
Thin (−50%)	0.008*	0.008*	0.008*	0.117	0.460	0.838	0.086
Thick places (+50%)	0.000*	0.000*	0.000*	0.001*	0.087	0.700	0.005*
Nep values (+280%)	0.000*	0.000*	0.000*	0.034*	0.011*	0.780	0.120
Yarn hairiness (H)	0.000*	0.000*	0.000*	0.000*	0.000*	0.001*	0.001*
Yarn tenacity (cN/tex)	0.874	0.020*	0.002*	0.028*	0.000*	0.788	0.009*
Elongation at break (%)	0.013*	0.001*	0.321	0.026*	0.225	0.331	0.042*
Fabric thickness (mm)	0.000*	0.000*	0.000*	0.063	0.383	0.488	0.039*
Bursting strength (kPa)	0.014*	0.237	0.030*	0.035*	0.021*	0.194	0.029*
Air permeability ($l/m^2/s$)	0.000*	0.000*	0.000*	0.014*	0.200	0.001*	0.472

**Significant for α = 0.05.*

Table 4.
The results of variance analysis of the effects of different waste types and waste ratio on yarn and fabric properties.

Fabric pilling values show that blowroom and fabric wastes containing fabrics have more pilling. Fabrics produced with card waste have a lower pilling degree and higher short fiber content but lower short fiber length. Probably, short fibers are too short to compose a pill in card waste.

4. Conclusions

In this study, the waste of yarn spinning mill and recycled cotton fiber from fabric waste were blended different blending ratios (10, 30, and 50%) with virgin cotton fiber. Ne 20/1 cotton open-end yarns were produced from these blends. The yarn physical properties of these yarns were tested. The effects of waste ratio and waste type on yarn properties were investigated. According to the test results, the dirty waste (blowroom waste and card waste) used in this study showed a different tendency from the clean waste (sliver waste and fabric waste fiber). The blowroom and card waste occurred during the cleaning of the fibers. They contain too much short fibers and vegetable matters. Especially, card waste contains the shortest fibers used in this study. For this reason, the yarn hairiness values are greater than the others. Besides, the yarn tenacity containing 50% card waste was lower than the others. However, in case of the use of blowroom waste, the number of thick places and the number of nep were found greater than the others. On the other hand, the content of the blowroom waste in the yarns has not influenced the yarn properties except for thick places and nep values up to ratio of 30%. When fibers recycled from the fabric have been used, it has been seen that the results are similar to those obtained when using the blowroom waste.

Fabric thickness, bursting strength, air permeability, and pilling values of fabrics were tested. When the results are examined, the use of blowroom waste up to 30% in the fabric has not affected the bursting strength adversely. However, when the ratio increased to 50%, the bursting strength decreased. When card waste was used, the values of bursting strength decreased as the waste rate increased. For the bursting strength values, it can be said that it is possible to blend up to 50% of the cotton fibers recycled from fabrics. Depending on the yarn hairiness values, the air permeability values of fabrics decreased as the ratio of waste fibers increased, except for the sliver waste.

Today, the recovery of waste and its use in specific proportions within the yarn blend is increasing day by day. According to the results of this study, it can be said that the blowroom waste and the fiber recycled from the fabric gave similar results and that the use of the wastes up to 50% did not affect the yarn and fabric properties adversely. It caused greater results in terms of only thick places and nep values than the other ones. On the other hand, card waste can be used in yarn blends up to 30%, depending on the high short fiber content they contain. Sliver wastes showed similar results compared to 100% virgin cotton fibers, as they are already clean wastes.

According to test results of the pilling values of fabrics, the best values were obtained when the card waste is used. On the other hand, the card waste has the highest short fibers content. It is thought that the short fibers are easily removed from the fabric surface after forming the pills in the fabric by itself and thus do not cause deterioration in appearance.

Acknowledgements

This research is supported by a project (Project no: 15-TKAUM-002) from Ege University Textile and Apparel Research and Application Center.

Photos of wastes and machines in **Figure 2** were taken by the authors in a Turkish spinning mill.

Author details

Tuba Bedez Ute[1], Pınar Celik[1*] and Memik Bunyamin Uzumcu[2]

1 Department of Textile Engineering, Ege University, Izmir, Turkey

2 Department of Textile Engineering, Bartın University, Bartin, Turkey

*Address all correspondence to: pinar.celik@ege.edu.tr

References

[1] Pickin J, Randell P. Australian National Waste Report 2016. Victoria, Australia: Department of the Environment and Energy & Blue Environment Pty Ltd.; 2017

[2] Murthy HVS. Introduction to Textile Fibers. New Delhi, India: Woodhead Publishing India Pvt. Ltd; 2016. 238 p

[3] Preferred Fiber; Materials Market Report 2018. 2018. Available from: http://textileexchange.org/wp-content/uploads/2018/11/2018-Preferred-Fiber-Materials-Market-Report.pdf [Accessed: June 30, 2019]

[4] Gurudatt K, De P, Rakshit AK, et al. Spinning fibers from poly(ethylene terephthalate) bottle-grade waste. Journal of Applied Polymer Science. 2003;**90**:3536-3545. DOI: 10.1002/app.12969

[5] Telli A, Babaarslan O. Usage of recycled cotton and polyester fibers for sustainable staple yarn technology. Tekstil ve Konfeksiyon. 2017;**27**:224-233. WOS: 000419066000003

[6] Mishra R, Behera BK, Militky J. 3D woven green composites from textile waste: Mechanical performance. Journal of the Textile Institute. 2014;**105**:460-466. DOI: 10.1080/00405000.2013.820865

[7] Briga-Sá A, Nascimento D, Teixeira N, et al. Textile waste as an alternative thermal insulation building material solution. Construction and Building Materials. 2013;**38**:155-160. DOI: 10.1016/j.conbuildmat.2012.08.037

[8] Binici H, Eken M, Dolaz M, et al. An environmentally friendly thermal insulation material from sunflower stalk, textile waste and stubble fibres. Construction and Building Materials. 2014;**51**:24-33. DOI: 10.1016/j.conbuildmat.2013.10.038

[9] El Wazna M, El Fatihi M, El Bouari A, et al. Thermo physical characterization of sustainable insulation materials made from textile waste. Journal of Building Engineering. 2017;**12**:196-201. DOI: 10.1016/j.jobe.2017.06.008

[10] Shukla SR, Harad AM, Jawale LS. Recycling of waste PET into useful textile auxiliaries. Waste Management. 2008;**28**:51-56. DOI: 10.1016/j.wasman.2006.11.002

[11] Karthik T, Gopalakrishnan D. Environmental analysis of textile value chain: An overview. In: Muthu SS, editor. Roadmap to Sustainable Textiles and Clothing. Singapore: Springer; 2014. pp. 153-188. DOI: 10.1007/978-981-287-110-7_6

[12] Pinheiro E, de Francisco A. Management and characterization of textile solid waste in a local productive arrangement. Fibres & Textiles in Eastern Europe. 2016;**24**:8-13. DOI: 10.5604/12303666.1201128

[13] Barker AF. Handbook of Textiles. Delhi: Abhishek Publications; 2009. 369 p

[14] Lord PR. Handbook of Yarn Production. Cambridge, England: Woodhead Publishing Limited; 2003. 493 p

[15] Agblevor FA, Cundiff JS, Mingle C, et al. Storage and characterization of cotton gin waste for ethanol production. Resources, Conservation and Recycling. 2006;**46**:198-216. DOI: 10.1016/j.resconrec.2005.07.002

[16] Lawrence CA. Fundamentals of Spun Yarn Tehnology. Vol. 509. CRC Press; 2003

[17] Halimi MT, Ben HM, Azzouz B, et al. Effect of cotton waste

and spinning parameters on rotor yarn quality. Journal of the Textile Institute. 2007;**98**:437-442. DOI: 10.1080/00405000701547649

[18] Halimi MT, Azzouz B, Ben Hassen M, et al. Influence of spinning parameters and recovered fibers from cotton waste on the uniformity and hairiness of rotor spun yarn. Journal of Engineered Fibers and Fabrics. 2009;**4**:36-44. DOI: 10.1177/155892500900400304

[19] Klein W, Stalder H. A Practical Guide to Opening and Carding. Textile Institute; 1987. 60 p

[20] Klein W. Blowroom and Carding, The Rieter Manual of Spinning. Vol. 2. Rieter Machine Works Ltd; 2014. 87 p

[21] Wulfhorst B. Technological and Economic-Aspects for The Waste Processing in Modern Cotton Spinning Plants. Melliand Textilberichte International Textile Reports. 1984;**65**:730. WOS: A1984TR92800013

[22] Duru PN, Babaarslan O. Determining an optimum opening roller speed for spinning polyester/ waste blend rotor yarns. Textile Research Journal. 2003;**73**:907-911. DOI: 10.1177/004051750307301010

[23] Hasani H, Semnani D, Tabatabaei S. Determining the optimum spinning conditions to produce the rotor yarns from cotton wastes. Industria Textilă. 2010;**61**(6):259-264

[24] Hasani H, Tabatabaei SA. Optimizing spinning variables to reduce the hairiness of rotor yarns produced from waste fibres collected from the ginning process. Fibres & Textiles in Eastern Europe. 2011;**86**:21-25

[25] Khan MKR, Rahman H. Study of effect of rotor speed, combing-roll speed and type of recycled waste on rotor yarn quality using response surface methodology. IOSR Journal of Polymer and Textile Engineering. 2015;**2**:2348-2181

[26] Khan KR, Hossain MM, Sarker RC. Statistical analyses and predicting the properties of cotton / waste blended open-end rotor yarn using Taguchi OA design. International Journal of Textile Research. 2015;**4**:27-35. DOI: 10.5923/j.textile.20150402.01

[27] Celep G, Doğan G, Yüksekkaya ME, et al. Geri Dönüşümlü Lifler İçeren Süprem Kumaşların Isıl Konfor Özelliklerinin İncelenmesi. Düzce Üniversitesi Bilim ve Teknoloji Dergisi. 2016;**4**:104-112

[28] Telli A, Babaarslan O. Commercialized denim fabric production with post-industrial and post-consumer wastes. Tekstil ve Konfeksiyon. 2016;**26**:213-220. WOS:000386121800013

[29] Yilmaz D, Yelkovan S, Tirak Y. Comparison of the effects of different cotton fibre wastes on different yarn types. Fibres & Textiles in Eastern Europe. 2017;**25**:19-30. DOI: 10.5604/01.3001.0010.2340

[30] Béchir W, Béchir A, Mohamed BH. Industrial cotton waste: Recycling, reclaimed fiber behavior and quality prediction of its blend. Tekstil ve Konfeksiyon. 2018;**28**:14-20. WOS:000440377600002

[31] Gun AD, Akturk HN, Macit AS, et al. Dimensional and physical properties of socks made from reclaimed fibre. Journal of the Textile Institute; 2014;**105**:1108-1117. DOI: 10.1080/00405000.2013.876152

[32] Yuksekkaya ME, Celep G, Dogan G, et al. A comparative study of physical properties of yarns and fabrics produced from virgin and recycled fibers. Journal of Engineered Fibers and Fabrics. 2016;**68**:68-76

[33] Yunus M, Rahman F. Micronaire effects. Textile Asia. 1990;**13**:58-61

[34] LENGTHCONTROL Fibre Length Tester. 2019. Available from: https://www.inteszt.hu/products/attachments/lengthcontrolpdf_qTHSvBouDR.pdf [Accessed: February 05, 2019]

Chapter 5

Phase Change Materials for Textile Application

Fabien Salaün

Abstract

The objective of this chapter is to determine which of the existing PCM families are more suitable for textile thermoregulation while proposing new solutions. Indeed, many of these materials are either limited by their overall enthalpy of phase change or by their thermal window. Thus, it focuses on the study of binary mixing allowing the widening of the temperature range of the phase change and the consolidation of the enthalpy balance by adding chemical species. PCM was microencapsulated to be applied onto textile substrate, before studying the thermal properties.

Keywords: phase change material, textile thermal comfort, paraffin, thermal transition, thermoregulation

1. Introduction

The French textile and clothing industry is facing many difficulties related to competition from countries with low labor costs and relocation. To cope with this, the players in the sector are adapting and diversifying their offer. Traditional industries and markets being the most affected, they are thus moving towards product niches, such as technical and high value-added textile markets, where know-how and innovation play a significant role. New products within particular advanced functionality regarding thermal comfort can help to position themselves in innovative product niches and meet consumer expectations.

The textile structure is involved in maintaining the thermal balance between the heat produced by the body and that given to the external environment. When adequately adapted, it protects the human body from hyper and hypothermia. Its resistance to heat and moisture transfer characterize it. However, its characteristics depend on external factors such as speed and relative humidity. Thermal insulation in textiles is traditionally made from voluminous composite materials where trapped air plays a key role. Since the late 1980s, a new type of insulation material has appeared on the market, phase change materials (PCMs). Integrated into clothing, they can improve their thermal behavior, regardless of their thickness, guaranteeing a greater freedom of movement for the wearer.

Phase change materials have been used for various applications in thermal energy storage since the 18th century, and many commercial products have been developed to this end. Depending on the field of application, PCMs are selected according to their temperature and phase transition enthalpy. Thus, when their phase change temperatures are relatively low (between −20 and −10°C), these materials are more likely to be used for food storage, while for temperatures between 2 and 15°C, they are generally used for air conditioning, that is, comfort applications [1]. Higher transition temperatures allow these materials to be used for solar energy storage [2], in agriculture [3], electronic equipment protection [4], or textiles [5–7]. Thermal energy storage by solid-liquid phase change has been the subject of a census by Zalba on more than 150 existing PCMs, 45 of which are commercially available [8].

The thermal properties of phase-change materials allow them to be perceived as the material of choice for thermal insulation of the human body. The possibility of keeping the wearer as long as possible in his thermal comfort zone, and at the same time to reduce the thickness of the garment, is a conceivable objective, taking into account the thermal insulation capacity of textile support containing PCMs. Indeed, these are active during the phase change period and stop when the phase change of all PCMs is complete. This insulating effect is generally referred to as effective thermal insulation. The choice of PCMs to be used is therefore based on the feeling of comfort that the user can feel, regardless of his metabolic activity and external conditions. The effectiveness of PCMs inserted in a textile structure will depend mainly on the temperature differential between the body temperature and that of the surrounding environment. Therefore, a product ideally having a thermal window from 19°C (vasoconstriction temperature) to 37°C should contribute to localized thermal regulation.

The conventional phase change material formulations have some disadvantages related to their chemical structure [9], justifying that they cannot be used without being contained in a capsule or trapped by capillarity in a graphite matrix [10] or chemical gel [11]. Moreover, since these materials can be in a liquid state, they cannot easily be incorporated into a textile carrier without being contained in capsules. Besides, they must be as small as possible in order to facilitate their integration and thermoregulation regarding heat exchange surface, thus helping to compensate for their low thermal conductivity. Regardless of the physical state of the microencapsulated material (solid, liquid, or both), it remains trapped inside. This allows it to be integrated into a textile coating and thus to keep its functionality as long as the coating remains intact.

Microcapsules incorporated in commercial textile composites contain only one active ingredient, usually paraffin. These systems are thus limited by the thermal properties, thermal window, and enthalpy of phase change, of the microencapsulated paraffin. The objective of this work is to develop new materials with improved thermal properties before their incorporation on textile support.

2. Thermal energy storage

The storage of energy in thermal form can be carried out according to three physical principles, that is, (i) sensible heat, (ii) phase change or latent heat, and (iii) thermochemical reactions.

2.1 Storage by sensible heat

Sensible heat storage is undoubtedly common and straightforward, but it is also very inefficient. A material (most often water, stones or thermal oil) is heated to a higher temperature whenever excess heat occurred and cooled when necessary. In general, volume storage capacities remain relatively low. Also, depending on the type of material selected, liquid or solid, the process will be limited in the first case by the chronic tendency to thermal destratification, and in the second by the low thermal conductivity associated with a high porosity rate. In these systems, there is always a need for a tank and also most often an exchange surface [12]. The cost price of these elements is generally the limiting factor in its economic application. The stored energy (Q in J g^{-1}) is dependent on both a temperature variation and a function of the mass of material (m) used and its calorific capacity C_p (Eq. (1)). The primary materials used for this type of storage are water, rocks, earth, and ceramics.

$$Q = m \int_{T_{initial}}^{T_{end}} C_p(T).dT \qquad (1)$$

2.2 Storage by chemical reactions or evaporation

Thermal energy storage by chemical reactions is achieved by using the energy released or required during a chemical reaction. The basic principle is as follows (Eq. (2)).

$$AB + Heat \rightleftharpoons A + B \qquad (2)$$

Using the heat provided by the external environment, the AB compound is divided into two components A and B, which can be stored separately. If A is mixed with B, the AB compound is reformed, and the heat is then released. The chemical reactions involved in this process must satisfy the total reversibility of the reaction since the products resulting from the reaction are likely to be separated and can be stored in a solid state and mixed to release heat when energy is required. Adding a catalyst can reduce the reaction temperature, but it is generally high. As a result, research on thermal energy storage by chemical reaction is still in the first stage and cannot be carried out in the short term. Nevertheless, other technologies have developed such as adsorption on activated carbon or zeolite, liquid phase absorption (LiBr), the reaction on chlorides ($MnCl_2$, $NiCl_2$) [13]. Storage systems based on chemical reactions have negligible energy losses, while sensitive heat storage absorbs stored heat from the environment and needs to be isolated from the outside environment [14].

2.3 Storage by latent heat or phase change

Storage by latent heat of fusion is carried out with little or no temperature variation since the phase change of a pure body is isothermal but by phase change of material. The case of alloys is different because melting takes place over a limited temperature range between $T_{solidus}$ and $T_{liquidus}$. Latent heat storage involves a first-order phase transition (enthalpy variation ΔH). When the transition is essential, the material is called PCM (Phase Change Material). These materials are compounds that can store and release thermal energy through their change of state, most often from solid to liquid, but also from solid to solid. When heated, the material takes calories from the external environment and reaches a temperature, T_{tr}, transition temperature, then passing from a phase 1 to a phase 2 by heat absorption. If cooled, the reverse transition occurs at T_{tr}; the material passes from phase 2 to phase 1 and returns all the previously stored energy to the external environment while remaining at the temperature T_{tr}. The energy involved is the endothermic or exothermic enthalpy variation of the phase change ΔH [15].

Since heat is closely linked to temperature (second principle of thermodynamics), this storage method is more interesting than the first because it allows energy to be stored at a given temperature or a given temperature level. Besides, storage is carried out at a small temperature difference, and it offers the possibility of restoring the stored energy at a constant temperature, at least as long as the solid and liquid phases are in equilibrium. The amounts of stored energy are also higher than when storing by sensible heat. It should be noted that in practice, not only the latent heat of state change but also the sensible heat of the liquid and the corresponding solid are exploited, which significantly increases the stored energy (Q) (Eq. (3)) [16].

$$Q = m.\int_{T_{initial}}^{T_{tr}} C_p(T).dT + m.\Delta H + m.\int_{T_{tr}}^{T_{end}} C_p(T).dT \tag{3}$$

The phase changes required for this type of storage are first-order transitions. However, only solid-solid and solid-liquid transitions are used [17]. Indeed, the liquid-gas transition is technologically unusable since it leads to too high a variation in volume, and the liquid-liquid transition is vigorously too weak to generate any interest. The use of the solid-solid transition is interesting only in the case where the transition is relatively energetic, that is, for plastic crystal-crystal transitions [18]. The absence of any liquid helps the use of the materials.

3. Properties of phase change materials

The use of latent heat from a phase change material is not in itself a new technique. At the end of the nineteenth century, to overcome the inconvenience of changing hot water bottles too often to heat railway cars, water was replaced by sodium acetate. This salt can store a large amount of heat which it releases entirely after a few hours. While a few patents were filed before 1973, it was not until the first oil shock that many specialized laboratories began research in this field. Taking into account the thermodynamic can only make the choice of good storage material, kinetic, chemical and economic criteria considered essential for the proper functioning of the application in question [19, 20].

3.1 Melting temperature: T_m

It is the first criterion for selecting a product, as it must be appropriate for the application. Its determination by differential calorimetric analysis is relatively easy. However, depending on the origin or purity (and the nature of the impurities) of the materials, it can vary by a few degrees. There are two types of fusion, one ideal called "congruent," the other called "incongruent." The ideal fusion is isothermal fusion, where when the material is at the melting temperature, the liquid and solid phases present are in equilibrium and have the same composition. In this case, the change of state occurs reversibly. Thus all the energy stored during melting is fully restored during crystallization, promoting the material's resistance during the cycles. This type of fusion is generally found in pure substances. The melting can also be incongruous. In this case, two phases are formed beforehand, one liquid and the other solid, and then a fully liquid phase is obtained. Generally, the material decomposes at a temperature below its melting temperature into liquid and crystals. A two-phase mixture with a solid compound of a different composition from the defined compound is obtained in the same way during the cooling. During the cycles, the initial material gradually changes, all the more quickly as variations in density can lead to phase segregation, which accentuates the degradation phenomenon.

3.2 The crystallization temperature/supercooling problem

The crystallization temperature does not necessarily coincide with that of the liquid to solid transition. There is sometimes a slight delay related to the crystallization kinetics which is dependent both on the crystal growth of the material and on nucleation phenomena. This phenomenon also called supercooling, corresponds to the fact that a body temporarily remains in a liquid state at a temperature below its crystallization point. This is a relatively common phenomenon since even water has a supercooling of a few degrees varying with its impurity level. From a microscopic point of view, supercooling is related to nucleation and the rate of growth of germs. The probability of germ formation is related to the creation of a solid/liquid interface requiring a certain amount of energy. The germ thus created must have a radius more significant than a critical radius, rc, in order not to be dissolved in the medium. The rate of growth of germs corresponds to the change in the size of the crystal over time. This velocity is zero at the melting point and increases rapidly as the temperature drops to a maximum value. Beyond this value, crystallization will depend mainly on the mobility of the molecules, linked to the viscosity of the medium. Thus, the growth rate decreases with increasing supercooling. Adding a nucleating agent, such as a saline hydrate, stable in the exploited thermal domain and chemically inert towards the storage medium, can eliminate this delay in crystallization. Another solution is to play on the roughness of the container walls to promote nucleation. Supercooling is also related to the mass of the product used; indeed, it decreases considerably with the use of large amounts. Thus, the results obtained by DSC, with masses in the order of mg, must be used with caution. In some cases, supercooling can be considered an asset, since it delays crystallization for a certain period of times.

3.3 The enthalpy of phase change

The enthalpy of fusion corresponds to the energy absorbed by the material during the solid/liquid transition. The liquid/solid transition is achieved by fully restoring this energy. Except in the case of significant supercooling, this is the enthalpy of crystallization. These enthalpies can be determined experimentally by DSC analysis, and like temperatures, differ by a few J g^{-1} depending on the purity and origin of the products. The literature considers that a suitable storage material must have a melting enthalpy, ΔH_m, higher than 180 MJ m^{-3}. Even if some reach 580 MJ m^{-3}, few are those to exceed 350 MJ m^{-3}. A change in volume accompanies a first-order transition, and for most materials, the volume expansion $\Delta V/V$ at fusion is positive except for ice and gallium and its alloys. Depending on the nature of the materials, it can vary from a few percent to 50%, and in particular from 5 to 15% for paraffin.

3.4 Thermal conductivity

Thermal conductivity, λ (W m^{-1} K^{-1}), is the amount of heat transferred in a unit of time through a material of a surface unit and a unit of thickness when the two opposite sides differ from a unit of temperature. For most materials, thermal conductivity decreases slightly as the temperature rises. Thus, in the liquid state, it is weaker than in the solid state. A high thermal conductivity minimizes temperature differences in the material during melting and crystallization, facilitating heat transfer. Different methods are possible to increase this conductivity, either by inserting fibers or metal matrices, graphite or urea or by increasing the exchange surface.

Thus, the choice of a suitable storage material can only be made by taking into account some of its intrinsic characteristics (melting and crystallization temperatures, enthalpy of phase change, volume expansion and thermal conductivity), but also by knowing its melting behavior and its ability to withstand the thermal cycle, not to mention economic constraints. The last criteria to be taken into account are chemical stability and its non-corrosive aspect to avoid storage problems and material-envelope compatibility.

4. Classification of phase change materials

Among the fusible materials likely to be suitable for heat storage by phase change, there are two main families, that is, (i) organic materials, and (ii) salt hydrates and their eutectics.

4.1 Organic PCMs

Among the organic materials used in latent heat storage, paraffin remains the most commonly used compared to fatty acids, polyethylene glycol (PEG) or benzene derivatives, which nevertheless have good characteristics (congruent fusion, excellent stability, good latent heat, and low corrosion). These fatty acids and their derivatives have excellent thermal characteristics. Nevertheless, their high cost is a barrier to their use. In the rest of this chapter, paraffin selected due to their high latent heat compared to other potential "choices" are mainly described. Paraffin is linear chain aliphatic hydrocarbons of the general formula C_nH_{2n+2}. Due to their molecular structure and complex structural and thermodynamic behavior, they have been the topic of a many number of studies for more than half a century. These compounds, which are widely distributed in nature in several forms, are found in gasoline and industrial solvents for some carbon atoms (n) between 4 and 10, or in diesel and fuel for $10 \leq n \leq 28$. At T = 295°K, the paraffin are gaseous ($1 \leq n \leq 5$), liquid ($5 < n \leq 15$), or solid ($n > 15$) form. On an industrial scale, these compounds are obtained either by cracking and isomerization or by fractional distillation of petroleum.

4.1.1 Chains conformation

The most stable conformation, for which the potential energy is minimal, corresponds to a zigzag arrangement or trans conformation. The symmetry of the molecule depends on the parity of the carbon number. Thus, even n-alkanes have an inversion center while odd-numbered ones have a plane of symmetry perpendicular to the chain. These molecules are likely to have conformational defects, varying according to chain length and temperature. These defects, illustrated in **Figure 1**, result from a sudden increase in the concentration of non-planar conformers among molecules. The first two defects, left terminal type defect and torsion type defect are the most frequent. Conformers with two left bonds are much less numerous, even if their concentration increases near the fusion. The concentration of defects increases significantly with temperature within each crystal phase. It is all the more marked when the temperature is close to the melting point of the materials.

4.1.2 Intermolecular interactions

Since molecules are composed only of carbon and hydrogen atoms, crystal cohesion is ensured through Van der Waals' forces (London, Keesom, Debye), and more specifically London's interactions.

4.1.3 Stacking of molecules

The general laws of compact molecular stacking were established by Kitajgorodskij in 1957 [21]. These laws obey two rules, maximum stacking, and free energy. During the trend towards maximum stacking, molecules minimize their energy by adopting compact arrangements that generate maximum contact. The molecular arrangements observed are those with the highest symmetry compatible with the structure of the molecules. In the case of aliphatic chains, the molecules

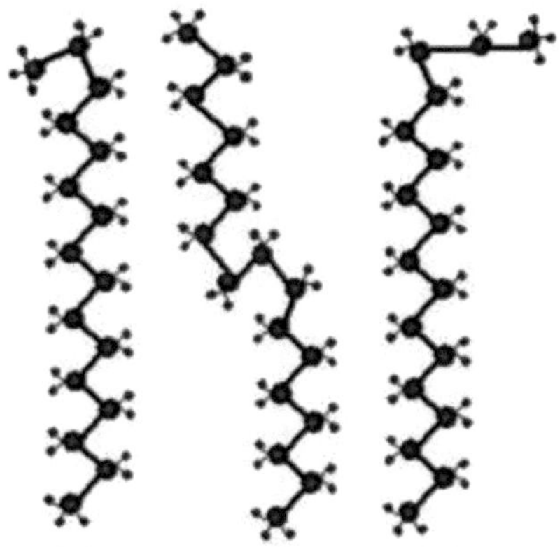

Figure 1.
Conformation defects, (a) left terminal type defect, (b) torsion type defect, and (c) two left connections defect.

are arranged in successive layers, in which the chains are parallel to each other. Therefore, the problem of stacking is at two levels, stacking within a layer and stacking these layers together. During his work, Kitajgorodskij also highlights three types of crystal arrangement: orthorhombic, monoclinic and triclinic.

4.1.4 Structural analysis

The phase changes required for energy storage correspond to first-order transitions. For practical reasons, only solid/solid and solid/liquid transitions can be used. Solid/solid transitions are rarely used because they are too slow, or the amount of energy involved is relatively small. Crystal energy results from two types of intermolecular interactions, those of methylene groups and terminal methyl groups. The values of these forces vary with the length of the chains, leading to differences in the stability of the crystal structure. Thus, n-alkanes crystallize under different structures and are polymorphic, that is, they are likely to undergo one or more solid-solid transitions before melting.

Each structure is part of an elementary mesh whose parameters are defined in the **Figure 2**, where $\vec{a}$, $\vec{b}$ and $\vec{c}$, are the basis vectors, and α, β, and γ, the angles between the Oxyz reference axes. The reproduction of this mesh a large number of times allows describing the whole crystal. Each mesh contains one or more molecules and has certain symmetry elements (symmetry axis, inversion center, mirror). A crystallographic group describes each structure:

- orthorhombic structure: $a \neq b \neq c, \alpha = \beta = \gamma = 90\ °$, noted β;
- triclinic structure: $a \neq b \neq c, \alpha \neq \beta \neq \gamma \neq 90\ °$, noted γ;
- monoclinic structure: $a \neq b \neq c, \alpha = \beta = 90\ °, \beta \neq 90\ °$, noted δ.

In the aliphatic chains of n-alkanes, each carbon atom is bound to two other carbon atoms and two hydrogen atoms, or one carbon atom and three hydrogen atoms at the end of the chains. The distance between the two carbon atoms is 0.153 nm, and the carbon-carbon binding angle is 114°; the latter value is higher than that of a perfect tetrahedron (109°8′). The most stable molecular arrangement in the solid state, which corresponds to the minimum in potential energy, is the one where the chains

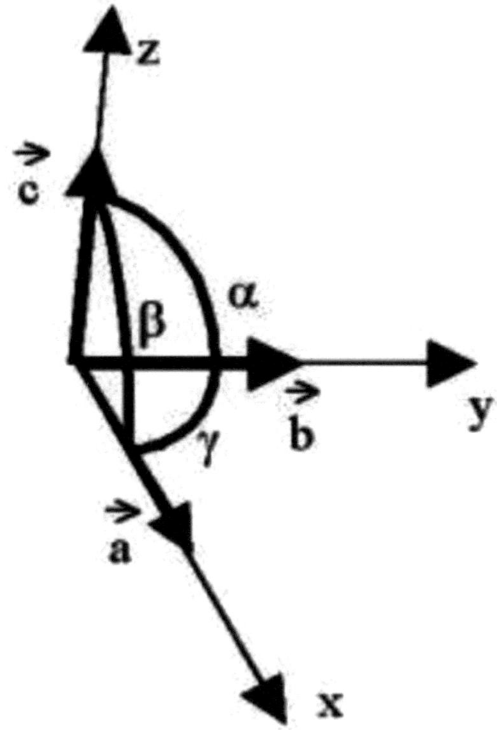

Figure 2.
Lattice constants.

adopt a trans or zigzag conformation. The structure is different according to the parity of the number n of carbon atoms of the n-alkane. Indeed, the steric congestion conditions imply that the CH_2—CH_3 bonds on either side of the stacking plane are aligned, regardless of the parity of the number of carbon in the n-alkane chain.

4.1.4.1 Low-temperature crystalline structures (ambient temperature)

Historically, the structure of n-alkanes has been discovered by Miller, based on the structure of n-nonacosane (n-$C_{29}H_{60}$) [22]. Terminal methyl groups are arranged in planes forming parallel or inclined lamellar surfaces to the axis of the molecular chains. In the solid state, n-paraffin has a much more compact molecular arrangement than in the liquid state. Some paraffin compounds are polymorphic in a defined range of pressure and temperature; they crystallize in different forms. The crystalline phases can be distinguished, in which the chains of carbon atoms are in a perpendicular position concerning the planes of the terminal methyl groups, relative to the orthorhombic structure, noted β_0 (n-C_{2p+1}); and inclined for the planes of the terminal methyl groups, relative to triclinic and monoclinic noted γ_0 (n-C_{2p}) and δ_0 (n-C_{2p}), respectively [23].

The orthorhombic structure β_0 (n-C_{2p+1}) found in odd alkanes ($n \geq 5$) includes four molecules per mesh, two layers of molecules that generate the periodicity in a direction perpendicular to the stacking plane of the chains, with c = 2*L (L: length of a molecule), and the parallelism of the bonds on either side of the stacking plane of the molecules between the last group CH_2 and the terminal methyl group. The orthorhombic structures of n-alkane mixtures are isomorphic to those of pure n-alkanes. However, they have irregularities at the end of the chain, caused by differences in the length of the molecules.

The series of even n-alkanes between hexane and hexacosane ($6 \leq n \leq 26$) adopts a triclinic structure (γ_0 (n-C_{2p})) with a space group P and some patterns per mesh equal to 1. The series of even n-alkanes ($n > 26$) has a monoclinic structure with two molecules per cell. Besides, it can be noted that the axis of the molecule is inclined to the plane of the terminal methyls. The periodicity is in an oblique direction with regards to the plane by considering the alignment of the links on each side of this plane. Thus, even n-alkanes crystallize in triclinic or monoclinic structures (δ_0 (n-C_{2p})) [23]. **Figure 3** shows this type of molecular arrangement in a triclinic structure along the axis of the chains, as well as the projection on the perpendicular plane.

4.1.4.2 Structure of the high-temperature solid phases (slightly lower than the melting temperature)

High-temperature solid phases called "rotator phases" exist only in a few degrees between the transition temperature in the solid state and the melting temperature. These high-temperature phases maintain an organized structure. In these rotator phases noted, respectively, β-RI and α-RII, the disorder is caused by a 180° rotation of some chains around their axis, which allows them to take a balanced position [24].

The β_0 (n-C_{2p+1}) low-temperature phases of odd n-alkanes undergo a solid state transition accompanied by a significant enthalpic effect that characterizes the appearance of a new orthorhombic phase, noted b. Then, this phase passes into a state called rotator noted β-RI, which has been highlighted by heating only in odd n-paraffin for $9 \leq n \leq 25$. This phase has an orthorhombic structure with the space group Fmmm, where two layers of molecules generate the periodicity in the direction perpendicular to the stacking plane according to the sequence ...ABAB... In the RI state of the β phase, the parameters (a, b) of the mesh continuously change without changing the space group: this causes an evolution as a function of temperature, which certainly reflects a second-order transition.

The rotator phase noted, α-RII, was highlighted in the case of even and odd paraffin for $22 \leq n \leq 26$. The increasing temperature study of the β-RI phase in odd n-alkanes (tricosane and pentacosane) identified the α-RII phase. The transition from the b-RI phase to the α-RII phase occurs when the mesh parameter ratio b/a is equal to $\sqrt{3}$. A change in the stacking sequence accompanies this evolution with the succession ... ABCABC... for the phase α-RII. The transition from the β-RI phase to the α-RII phase is a first-order transition, with a small enthalpy effect.

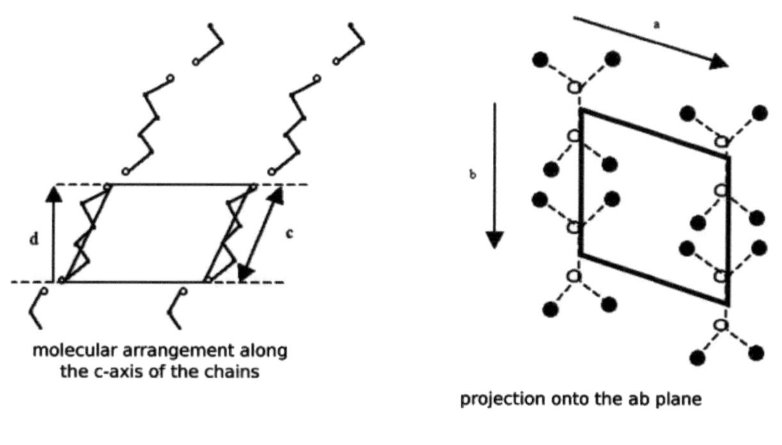

Figure 3.
Triclinic structure of even n-alkanes.

4.1.5 Thermodynamic analysis

Alkanes are generally characterized by their transition enthalpy (solid 1/solid 2) and melting enthalpy, noted H_{tr}, respectively, and ΔH_m, and by the temperatures at which these phase changes (T_{tr} and T_m) or state occur. We can also look at the global transition enthalpy, noted ΔH_g, which reflects the transition from solid state 1 to liquid (**Figure 4**).

4.1.5.1 Temperature and enthalpy of solid-solid transition

The phase transition temperature changes according to the length of the carbon chain and alternately according to the parity of the molecule. This alternation is because the final crystal structure is different according to the parity of the alkane. Indeed, an odd n-alkane will pass from the orthorhombic phase to the rotator I phase, while the even n-alkane (for $n \geq 20$) will pass from a triclinic or monoclinic phase to a rotator II phase. This alternation is also found on transition enthalpies from $n = 20$; indeed, it appears that the enthalpy is lower for odd n-alkanes than for peers, the difference can be attributed to the evolution for odd ones from the β-RI phase to the α-RII phase (**Figure 5**).

The notation ΔH^* corresponds to the summation of the transition enthalpy (β-RI/α-RII) and the increase of enthalpy in the β-RI phase. This increase in enthalpy is related to the calorific capacity of this relatively large phase. Thus, this enthalpy gain in the β-RI phase contributes to the even-odd alternation. Nevertheless, solid-solid transitions are generally not very energetic, so they are not always visible in calorimetry.

4.1.5.2 Melting temperature and enthalpy

According to the parity of the n-alkane, and according to the number of carbon atoms it contains, it can be observed an even-odd alternation on the two melting temperature evolutions. This disappears when the structures before fusion of the even and odd n-alkanes are identical, that is, for $n > 20$, as a consequence of the structural difference between the triclinic and rotator I phases.

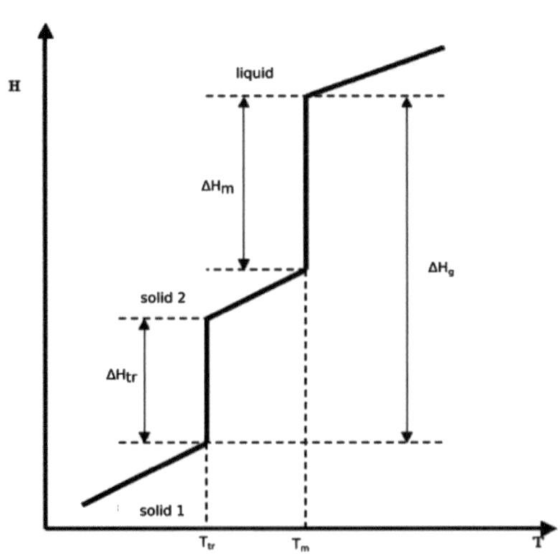

Figure 4.
Evolution of the enthalpy of a pure n-alkane as a function of temperature.

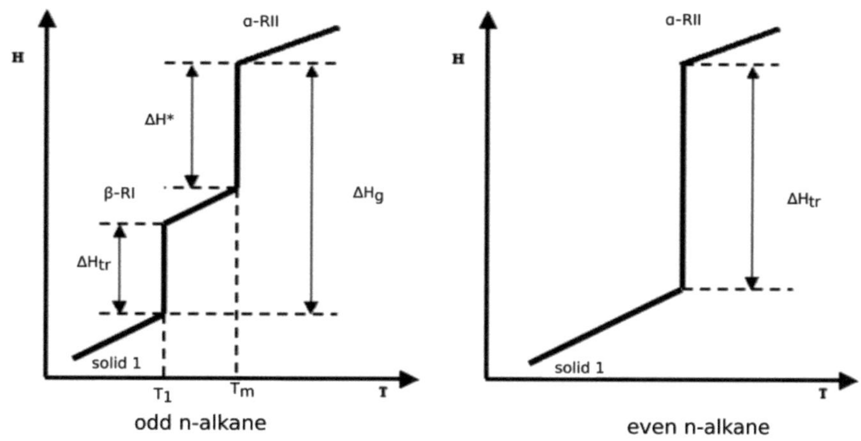

Figure 5.
Transition enthalpy of odd and even n-alkanes.

4.1.5.3 Global enthalpy

The global enthalpy represents the enthalpy balance of all transformations from the solid state to fusion. Its variation is practically linear as a function of the number of carbon atoms [25]. It, therefore, seems that this balance is independent of the nature of the low-temperature phase and the parity of the n-alkane.

4.1.6 Improvement of thermal conductivity

Much research has attempted to compensate for the poor thermal conductivity of paraffin by adding fins or expanded metal to the material or by dispersing it in a porous conductive material such as natural expanded graphite [26]. This method makes it possible to obtain composites with high thermal conductivity, high paraffin mass content (65–95%) trapped by capillarity [27]. A more common solution is to divide paraffin by volume by encapsulating it in PE spheres or by dispersing it as an emulsion in water. However, such a solution has the effect of considerably reducing (50%) the effective volumetric heat capacity of the process. In other words, its power is improved, but it takes up twice as much space.

4.1.7 Binary systems of n-alkanes

By combining alkanes in pairs, it is possible to develop systems of various structural types, some of which have very interesting thermal characteristics for energy storage [28]. The study of the behavior of binary mixtures of n-alkanes through phase diagrams or phase equilibrium has generated growing interest in the design of latent heat storage materials since it allows the prediction of the system's behavior during phase changes. Nevertheless, this behavior is governed by some general rules of thermodynamics on the possibility of forming solid solutions.

4.1.7.1 Gibbs rule

The solid solution of two n-alkanes is favored if the Gibbs free energy of the mixed crystal is lower than that of each of the pure constituents; otherwise, there is no miscibility in the solid state, but a eutectic.

4.1.7.2 Symmetry rule

Two n-alkanes are only miscible if they have the same crystal structure. This prerequisite, which was first formulated by Kitajgorodskij, is necessary but not sufficient [21], concerns n-alkanes of neighboring chain lengths.

4.1.7.3 Kravchenko rule

Kravchenko defines a criterion that predicts the mutual solid state solubility of two n-alkanes in the form (Eq. (4)) [29]:

$$\tau = \frac{n - n'}{n} \tag{4}$$

With n and n′ denoting the number of carbon atoms contained in each molecule. Thus, the total solubility is obtained for $\tau < 0.06$, it is partial when $0.06 < \tau < 0.15$, and almost null for $\tau > 0.15$.

However, the latter rule is insufficient since the symmetry rule implies that n-alkanes below C_{36} whose chain length differs from one unit cannot have total solubility since they are not isomorphic. Thus, many authors present phase diagrams in the literature where this rule is not respected. Also, Agafonov et al. [30], and He and Setterwall [31], also reported that the mixture of two isomorphic components in the solid state forms a continuous solid solution.

Therefore, synthetic paraffin has many advantages (no or low subcooling, perfect thermal reversibility) but remain relatively expensive and have a very low thermal conductivity ($\lambda = 0.24\ W\ m^{-1}\ K^{-1}$), which inhibits phase change kinetics and reduces the available powers accordingly. Their latent heat and density increase as the number of carbon atoms increases from C_1 to C_{40} and stabilizes around 280 J g^{-1} and 820 J m^3, respectively. The melting temperature also varies very gradually from 91 to 388°K depending on the number of carbon atoms, which potentially makes the use of this family of materials very flexible.

4.2 Inorganic PCMs

Inorganic PCMs, and more specifically saline hydrates, represent an interesting alternative to paraffin because of their low cost, high storage density (between 250 and 400 MJ m^{-3}), and relatively good thermal conductivity [32]. However, the main problem encountered when using these materials is phase segregation during melting [33]. Indeed, many of them have an incongruous fusion, making the system irreversible.

4.2.1 Physico-chemical properties of materials

The realization of energy storage by latent heat is related to the physico-chemical properties of the materials used, that is, (i) subcooling, (ii) crystallization rate, (ii) and the melting. A rapid rate of crystallization is generally required to allow the energy absorbed by the material to be released within a reasonable time. However, latent heat storage systems, because of their operation under a low-temperature difference, require a low crystallization rate. Acceleration of crystallization kinetics can be achieved by introducing solvents with high polarity and a high dielectric constant in order to improve ion mobility. Hydrated salt can be classified according to their type of melting into four categories, that is, (i) congruent melting, (ii) semi-congruent melting hydrates, (iii) non-congruent melting, and

(iv) eutectics [34]. Of these four types of transitions, only two correspond to perfectly reversible transitions: congruent melting and eutectic. The other two are characterized by the transition from a solid single-phase state to a liquid/solid two-phase state. The degree of incongruous melting varies according to the hydrates considered. Nevertheless, some hydrates do not show incongruent fusion, but their cost is relatively high.

4.2.2 Solid-liquid balance

During its fusion, the salt, noted $X(Y)_n{\cdot}mH_2O$, is likely to undergo one of the following two reactions (Eqs. (5) and (6)).

$$X(Y)_n mH_2O \rightarrow X(Y)_n . kH_2O + (m - k) H_2O \quad (5)$$

$$X(Y)_n mH_2O \rightarrow X(Y)_n + mH_2O \quad (6)$$

Hydrated salt tends to separate into two phases, no longer maintaining the salt/water ratio in suitable proportions. A salt-rich liquid (denser) tends to accumulate it at the bottom of the container, and a water-rich salt solution floats over the whole. When repeating fusion/crystallization cycles, less and less salt and water can come together and react to form a hydrate with the desired melting temperature. Part of the solid cannot be melted, and part of the liquid cannot be crystallized. The effect is a gradual loss of phase change enthalpy and latent heat storage capacity [35]. The most effective method to overcome this problem remains the suspension or agitation of the system. Suspension requires the introduction of clay or the formation of a gel that traps the hydrate so that when it crystallizes, salt and water are in contact to allow the system to recombine. The dimensions of the storage container can also be adjusted by minimizing them to limit phase separation.

4.2.3 Anhydrous rearrangement

The effectiveness of these materials is related to the amount of water and salt that can react to form the hydrated form. In the case of sodium sulfate, it appears that the salt settles very quickly, and rearranges [36]. As a result, it only partially hydrates, generating a low storage material. The addition of a continuous quantity of water up to the crystallization point makes it possible to overcome this problem. Nevertheless, shaking the system is strongly required to optimize the formulation of the PCM.

Therefore, among inorganic compounds, only saline hydrates and their eutectics have acceptable properties, including high latent heat and relatively low prices. Nevertheless, these materials have some limitations for latent heat energy storage including subcooling and their tendency to melt incongruently. This phenomenon is all the more embarrassing as it leads to the non-reversibility of the change of state. This problem can be eliminated by adding a clay-based gelling mixture that inhibits the settling effect of the lower hydrate [37]. Reversibility is thus restored within a reasonable time. The final limitation for these compounds is their low crystallization rate, which limits the thermal storage capacity. Some solvents or additives with high polarity and high dielectric constant accelerate the kinetics [38]. Indeed, better mobility of ions and therefore a high dielectric constant leads to a decrease in liquid-solid interfacial energies.

The performance of a latent heat storage system is based on the thermal properties of the compounds selected according to their melting and crystallization temperature, which must be in adequacy with the intended application, but also the enthalpy of phase change, and their stability during their implementation. The choice of one family over another is generally a function of the application and the initial cost of the products.

5. Formulation for textile thermoregulation

In recent years, phase-change materials have generated particular interest in thermal energy storage. The advantage of using latent heat storage lies in the possibility of optimizing the thermal windows of use (or "activation") by the judicious choice of materials based on both temperatures and phase transformation enthalpies. The action of PCMs incorporated in textile composites is "temporary" or "transient," that is, it is effective as a barrier to thermal energy until all latent heat stabilizing the exchange temperature is absorbed or released during the phase change of the material. Thus, by choosing a PCM formulation adapted to exchanges, thermal energy can be stored or released, and can effectively be "recharged" by heat or cold source. Among the existing PCMs, we have chosen to work with paraffin, or n-alkanes, which are ideal candidates for latent heat thermal energy storage due to their thermal characteristics with phase change enthalpies in the order of 200 J g^{-1} and phase transition temperatures varying according to their number of carbon atoms in the molecules (**Table 1**).

The objective of this part is first to identify potential candidates from among all present n-alkanes likely to be suitable for textile thermoregulation, then to develop a formulation by binary mixing, by determining thermo-physical properties and by energy characterizations by differential enthalpy analyses.

5.1 Choice of materials: enthalpy analysis of pure substances

The thermal characteristics of the selected n-alkanes (n-hexadecane, n-heptadecane, n-octadecane, n-nonadecane, and n-eicosane) were studied by differential calorimetry using the computer-controlled TA Instrument type DSC 2920 apparatus using the TA Advantage control software. The analyses are carried out under nitrogen flow with a flow rate of 50 ml min^{-1}. The samples, with a mass of about 4.0 ± 0.1 mg, are placed in an aluminum crucible, closed by a cover on which two holes have been made to ensure the chemical inertia of the medium.

5.1.1 The even n-alkanes

The thermograms of n-hexadecane (**Figure 6**) and octadecane (**Figure 7**), during a fusion-crystallization cycle, are characterized by an endothermic peak during the heating and transition from triclinic to liquid structure and an exothermic peak during the cooling (liquid to triclinic).

The enthalpies and phase change temperatures measured are in accordance with those found in the literature. In both cases, it is also observed that the melting peak is always preceded by a slight deviation from the baseline a few degrees before the melting, due to the existence of pre-melting phenomena.

The case of n-eicosane is different. Although the 2°C min^{-1} cycles initially allowed us to measure enthalpies and phase change temperatures similar to those cited in the tables, the exothermic peak raised some questions. Indeed, the thermogram (**Figure 8(a)**) shows that this compound does not have any polymorphism during the temperature rise while the cooling presents a cluster of peaks that deconvolves into two distinct peaks during a cycle at 0.5°C min^{-1} (**Figure 8(b)**). We observe the liquid/a-Rotator II/triclinic transitions. The existence of this

n-alkane	Carbone atoms	Melting point (°C)
n-octacosane	28	61.4
n-heptacosane	27	59.0
n-hexacosane	26	56.4
n-pentacosane	25	53.7
n-tetracosane	24	50.9
n-tricosane	23	47.6
n-docosane	22	44.4
n-heneicosane	21	40.5
n-eicosane	20	36.8
n-nonadecane	19	32.1
n-octadecane	18	28.2
n-heptadecane	17	22.0
n-hexadecane	16	18.2
n-pentadecane	15	10.0
n-tetradecane	14	5.9
n-tridecane	13	−5.5

Table 1.
Change in the melting temperature of n-alkanes as a function of the number of carbon atoms.

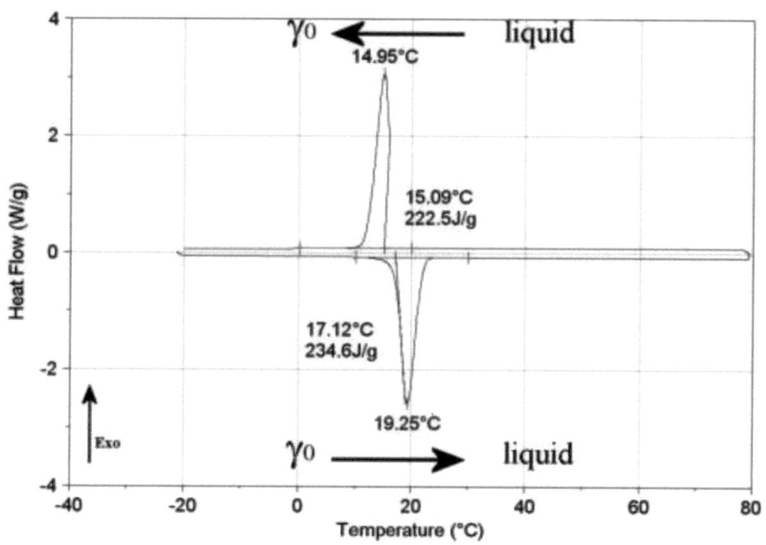

Figure 6.
Heating and cooling curves of n-hexadecane (N_2, 2°C min^{-1}).

rotating structure varies between 30 and 34°C, depending on the number of cycles imposed on the sample.

During his work, Espeau also noticed this type of variation for different samples analyzed under the same conditions [39]. The existence of this phase was not observed when the temperature increased. To observe it, it would have been necessary to raise the temperature after crystallization and before the solid-solid phase

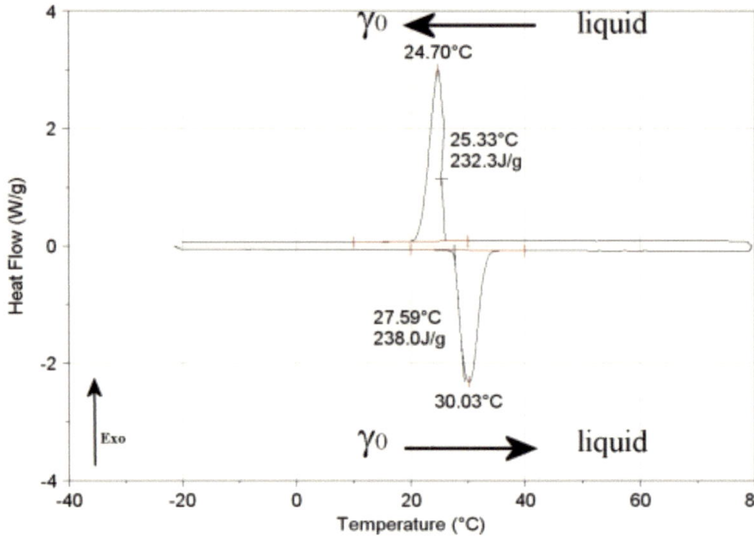

Figure 7.
Heating and cooling curves of n-octadecane (N_2, 2°C min^{-1}).

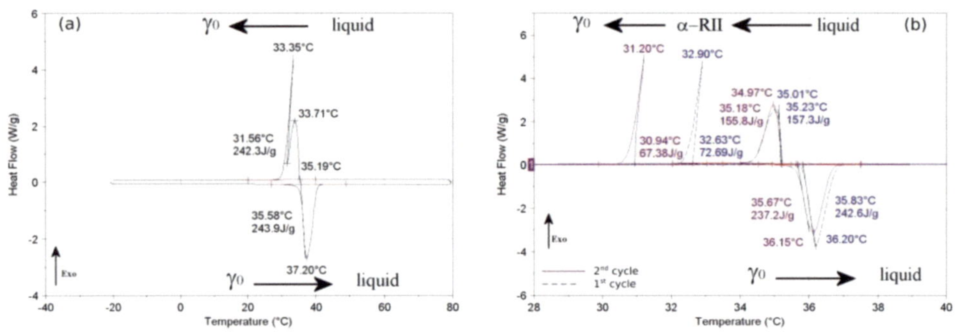

Figure 8.
Heating and cooling curves of n-eicosane (N_2, 2°C min^{-1} (a) and 0.5°C min^{-1} (b)).

change. The solid-solid transition enthalpy is slightly lower than that measured by Espeau. This difference can be blamed on the measuring instrument. Indeed, it was noted that whatever the temperature ramp imposed for this type of paraffin, the DSC did not keep its set point during the phase change. This phenomenon can be minimized by reducing the mass of the sample, but can still be observed for a ramp of 0.5°C min^{-1} and a mass of about 1 mg.

5.1.2 The odd n-alkanes

The thermograms presented in the **Figures 9** and **10** show the existence of two endothermic phenomena related to the solid phase transition of the orthorhombic structure, b0, to the rotator I phase, and to the solid-liquid transition by the rotator I/liquid fusion. Similarly, the temperature decrease is marked by the presence of two exothermic peaks corresponding, respectively, to the liquid/rotator phase change I and the solid/solid rotator transition I/β0. The temperatures and enthalpies of phase changes and solid/solid transitions are of the same order of magnitude as those measured by Espeau but slightly lower than those of Barbillon's work [40].

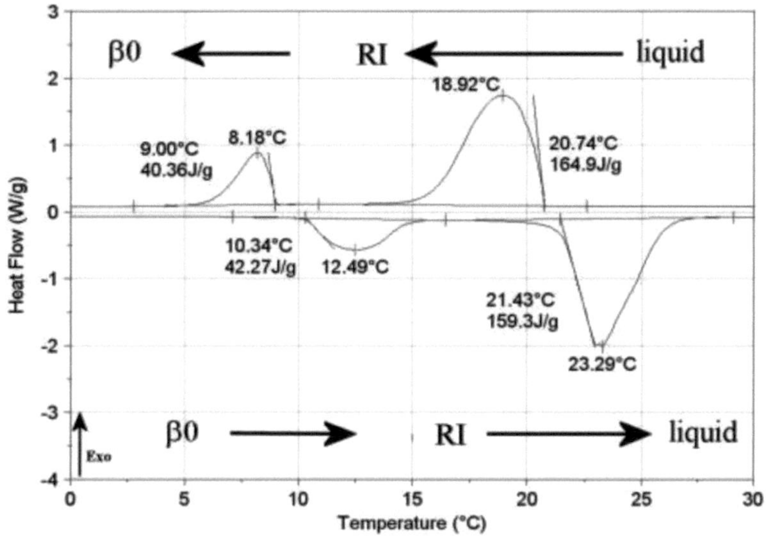

Figure 9.
Heating and cooling curves of n-heptadecane (N_2, 2°C min^{-1}).

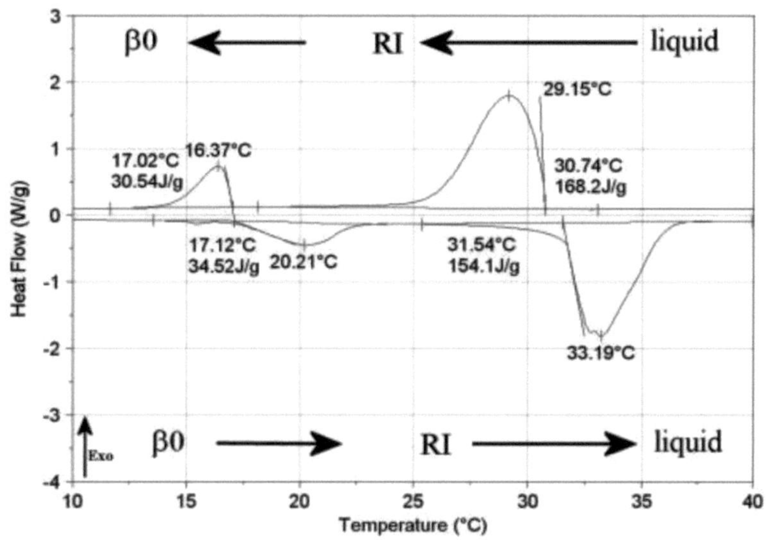

Figure 10.
Heating and cooling curves of n-nonadecane (N_2, 2°C min^{-1}).

6. Thermal span adjustment of binary mixture C_{16}/C_{20}

The study of simple compound made it possible to thermally characterize the different kinds of paraffin likely to be suitable for textile thermoregulation. However, none of them have a sufficiently wide thermal span between 19 and 30°C. Odd n-alkanes appear to be of little interest given the existence of a solid/solid transition with low energy and a lower enthalpy of solid/liquid phase change than for even n-alkanes. Also, their cost is four times higher than that of even n-alkanes, justifying the fact that these two compounds do not appear to be ideal candidates for this research. Of the remaining three n-alkanes, we focused on the binary mixture C_{16}/C_{20}, due to their respective melting temperatures on either side of those required for the application. Also, the objective is to obtain a succession of phase transitions, which can be induced by the superposition of crystals of different sizes and the partial miscibility of n-alkanes. It is hoped that all transitions will be

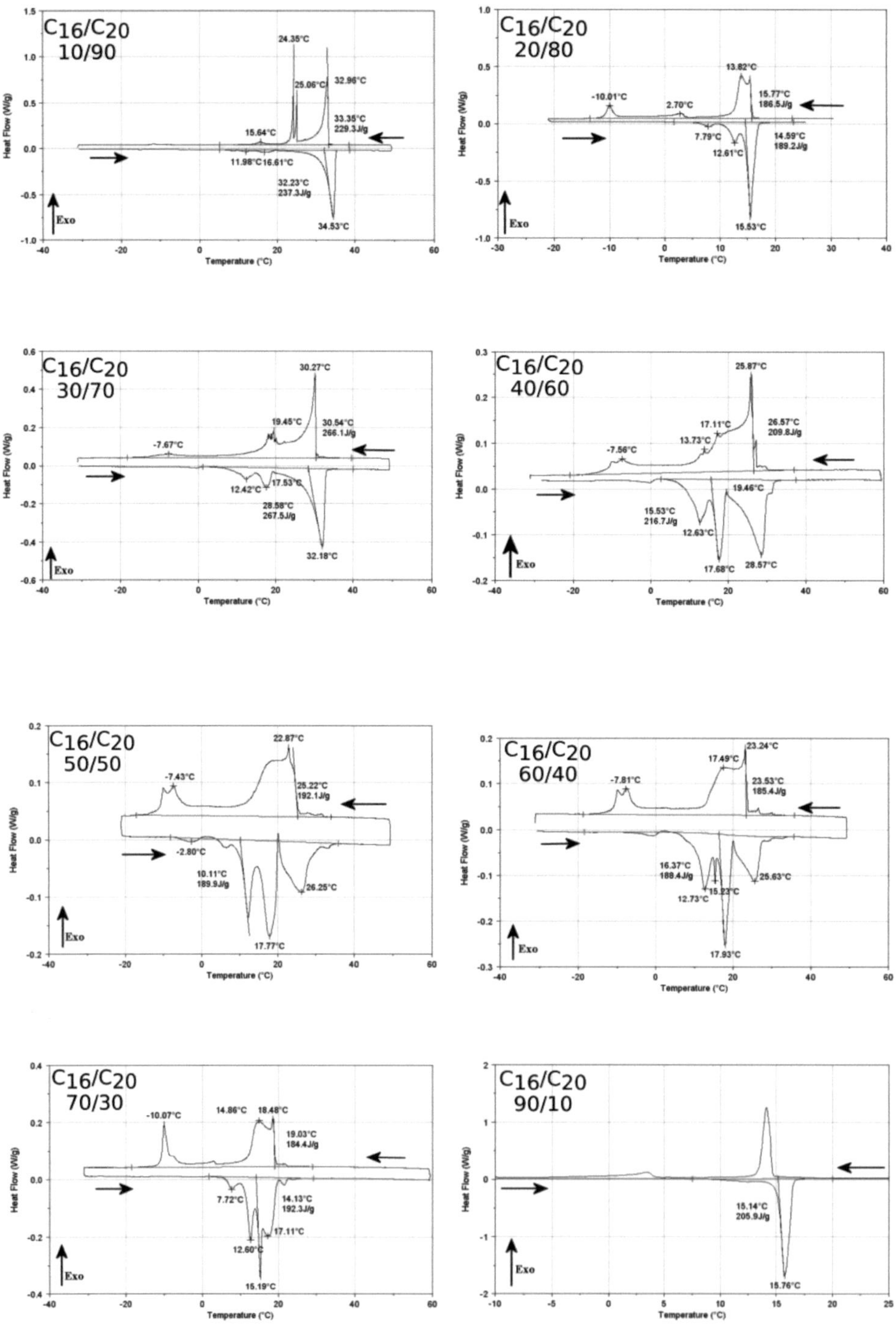

Figure 11.
Heating and cooling curves of C16/C20 binary mixture at various ratio.

contained within a single, relatively large peak. The binary diagrams C_{16}/C_{18} and C_{18}/C_{20} presented in Métivaud's work show that solid/solid and solid/liquid transitions occur over relatively narrow and distinct temperature ranges and are therefore not suitable for textile application [41–43].

The enthalpy characterization of the C_{16}/C_{20} mixture in different proportions was performed with 3 mg samples and a temperature ramp of 0.5°C/min, allowing the peaks for the different transitions to be dissociated. All the thermograms relating to this mixture are presented in **Figure 11**. The superposition of the endotherms shows that when one of the compounds is in the majority in the mixture, the thermal phase transition windows are narrow and tend towards those of the alkane fusion in a higher proportion. On the other hand, for mass fractions between 0.3 and 0.7, there is an expansion of the endotherms between 0 and 35°C, implying the appearance of new solid/solid transitions within the material during the temperature rise.

The measurement of enthalpies shows that they vary between those of pure substances and 190 J g^{-1}, except in the particular case of mixing in a 30/70 ratio (hexadecane/eicosane). Thus, the widening of the thermal window is accompanied by a decrease of about 20% in the total enthalpy of phase changes. This "loss" is related to the increase in the number of solid/solid transitions that are less energetic than solid/liquid transitions. We finally chose the 50/50 mixture, which allowed us to load the material over a larger thermal window observed from 3 to 32°C for enthalpy of 190 J g^{-1}.

The binary mixture of the two kinds of paraffin disrupts the cohesion of the crystal, which results in a succession of solid/solid transitions inducing the widening of the thermal window and the decrease of the overall enthalpy. Based on this hypothesis, we focused on the introduction into the mixture of a "soluble" charge in any proportion whatsoever in hexadecane and eicosane. We chose tetraethyl orthosilicate. The measurements of the overall enthalpies of the mixture as a function of the charge ratio show that the latter makes it possible to increase the energy balance to values comparable to those of pure substances, then from 4% in weight, the enthalpy decreases until reaching its basic level at 20% (**Figure 12**).

Among the five n-alkanes selected, based on their melting temperatures, the even n-alkanes have a higher enthalpy of phase change and are less expensive. The binary mixing of n-hexadecane with n-eicosane allows the thermal window to widen with a decrease in phase transition enthalpies. The addition of a "miscible" charge in the pre-selected mixture consolidates the energy balance without modifying the thermal span.

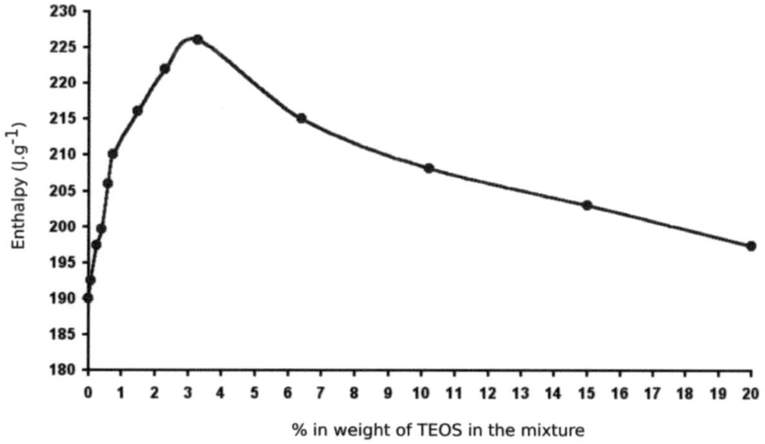

Figure 12.
Enthalpy variation vs. TEOS content.

7. Textile application

Since these materials may be in a liquid state, they cannot be easily incorporated into a textile carrier without being contained in capsules, which must be as small as possible to facilitate thermoregulation. Microencapsulation techniques may vary depending on the constitution of the membrane. Nevertheless, they all start with an oil-in-water or water-in-oil emulsion step depending on the solubility of the principle to be encapsulated in one or the other of these phases. In our case, the dispersion of paraffin in water is done by means of a rotor-stator. Microcapsule synthesis is continued by adding a melamine-formaldehyde-based prepolymer and reducing the pH for polycondensation of resin chains around PCM droplets. The formation of the aminoplast shell allows the active principle to be correctly isolated from the external environment and prevents any diffusion, and provides an interesting mechanical and thermal resistance for application on textiles. This in situ polymerization allows the production of microcapsules with a relatively narrow and controlled granulometry, by controlling the different physico-chemical parameters governing the emulsion stages (shear, pH, temperature, surfactants, interfacial tension...) and membrane growth [5, 6, 44].

Regardless of the physical state of the material inside these microcapsules (solid, liquid, or both), it remains trapped inside. This allows it to be integrated into a textile coating or incorporated into different artificial fibre compounds and can be effective as long as the coating or fibres themselves remain intact. For the application of these materials on textiles by bath impregnation, we used a thermodynamic approach to wetting to characterize the textile/binders/microcapsules interfaces. Thus, the comparison of the components of the surface energy of the polyurethane binders with those of the resins forming the membrane of the microcapsules, allowed us to optimize our formulation, before validating it by the ISO 6330 standard.

A 100% cotton fabric (566 dtex warp and 564 dtex weft yarns at densities of 26 ends/cm × 16 picks/cm, weighing 270 g m^{-2}, thickness of 0.50 mm), labelled COT, a 100% polyester fabric (345 dtex warp and 290 dtex weft yarns at densities of 18 ends/cm × ~7 picks cm^{-1}, weighing 178 g m^{-2}, thickness of 0.22 mm) (PES), and a PES nonwoven (155 g m^{-2}) were chosen as the textile specimens (**Table 2**).

The thermal resistance to exchanges is expressed in m^2 °C W^{-1}. The clothing isolation unit used is the clo, defined as the isolation of clothing necessary to maintain the thermal balance of a resting subject exposed to calm air and a temperature of 21°C. In practice, 1 clo corresponds to the isolation provided by classic streetwear and common underwear.

The heat transfer resistance for a sample, including the thin layer of air between the textile and the module, is calculated according to the Eq. (7).

$$R_t = \frac{(T_{sk} - T_a)A}{H} \tag{7}$$

where R_t is the thermal resistance in m^2 °C W^{-1}, A the surface area in m^2, T_{sk} the temperature (35°C), Ta the ambient temperature (°C) and H the heat flow in W.

The coating generates a modification of the surface and the context of the support, thus modifying the transfer of heat flows. Nevertheless, the evolution of R_t as a function of the mass deposited shows, and in particular in the case of cotton, that a low coating makes the textile less insulating and that a minimum quantity is required to reach the level of the reference sample. This characteristic is closely linked to the presence of an uniform or uneven deposition on the surface that modifies the contribution of air, by substituting free volumes with polymer volumes. Indeed, a low rate, as is the case for the COT1/2-(17) and PES1/2-(35) samples, does

Sample label	Ta = 21°C		Tb = 32°C	Weight deposited (g m^{-2})
	Rt (m^2 °C W^{-1})	Rt (clo)	Rt (m^2 °C W^{-1})	
Pristine cotton	0.070	0.451	0.094	0
COT1/2-(17)	0.068	0.438	0.085	17
COT1/2-(39)	0.069	0.445	0.090	39
Pristine PET	0.071	0.460	0.098	0
PES1/2-(35)	0.068	0.443	0.096	35
PES1/2-(50)	0.074	0.480	0.099	50
PES1/2-(101)	0.068	0.437	0.105	101
Pristine nonwoven	0.093	0.604	0.093	0
NT1/2-(65)	0.094	0.608	0.146	65
NT1/2-(95)	0.089	0.577	0.113	95

Table 2.
Thermal resistance of the obtained textile samples determined at 21 and 32°C.

not allow a uniform film to be obtained. Thus, there is a competitive phenomenon between the modification of factors influencing convection, such as the volume, porosity, and geometry of the textile, decreasing the amount of air within the material, and the action of microcapsules containing phase change materials. When the structure is uniformly covered, which is the case for PES1/2-(50) and PES1/2-(101), and that the film is sufficiently impermeable to air, thus to the renewal of the microclimate through the textile, the thermal resistance of the samples is higher than that of the reference.

The choice of a process for implementing microcapsules of PCMs on textile support remains a crucial step in the implementation of thermoregulatory structures. The functionalization of nonwoven is an excellent example since it is the structure with the highest % of free volume, its construction is random, and it is the thickest, that is, as much of factors that make the inherent insulating power of air predominate. Thus, whatever the mass deposited, its thermal resistance remains lower than that of the original textile at 21°C. The full bath coating with squeezing not only modifies the porosity of the material but also its compactness, making it difficult to exploit the results.

Therefore, the most suitable process for nonwovens is the lick roller coating. However, the thermal properties of the samples are worse than the reference but higher than those of NT1/2-(95). When analyzing their samples of foam-coated nonwovens and PCM microcapsules, Shim et al. [45] also, reveal that the thermal resistance of their materials reduced to one unit thickness is lower for structures containing PCMs than without.

The evaluation of the thermal behavior of textile structures has shown that it is linked to the deposition process used, the mass deposited, the nature and the context of the textile support used. The modification of the textile fabric structure during impregnations has a significant influence on the thermal behavior of textile composites. Indeed, small deposited masses generate a lower thermal resistance than the reference sample due to the decrease in air permeability. Also, the temperature gradient within the structure influences the loading of MPCMs. Thus, surface impregnation allows a higher rate of active ingredient to be activated over a wider thermal window. Because of the results, it is necessary to deposit a certain amount of MPCMs to compensate for the thermal insulation lost when the context is modified. Indeed, this modification during the coating with the binder and microcapsules reduces the porosity of the material. Thus, from higher mass deposits, the thermoregulatory effect of MPCMs is perceptible and allows to improve thermoregulation. Nevertheless, the optimization of the binder/microcapsule ratio is essential to more explicitly quantify the role played by these two elements in the perceived effect.

8. Conclusion

The use of PCMs is based on the encapsulation by an aminoplast membrane of a binary mixture of paraffin. Thus, as a first step, we determined the PCM adapted to our study, namely a binary formulation of n-hexadecane/n-eicosane (50/50), allowing the thermal window to be extended over a temperature range of −10 to 30°C with an overall phase change enthalpy of 190 J g^{-1}, that is, 20% lower than that of n-alkanes taken separately. The introduction of TEOS at 4% has made it possible to consolidate the energy balance at 230 J g^{-1}. In a second step, during the microencapsulation of this formulation with an aminoplast resin on a laboratory scale, we obtained microcapsules with an average diameter between 1 and 5 μm. We observed that the final particle size of the particles obtained was very dependent on the particle size resulting from the emulsification step.

The thermal characterization using a hot guarded plate showed the importance of the textile context during heat transfers. Thus, a minimum amount of binder and microcapsules deposited mass is required to compensate for the change in this context during the implementation step. This quantity is 20 g m^{-2} for binders and 40 g m^{-2} for microcapsules. Also, the responses to thermal variation depend on the mass deposited.

Author details

Fabien Salaün
ENSAIT-GEMTEX, Roubaix, France

*Address all correspondence to: fabien.salaun@ensait.fr

References

[1] Cabeza LF, Mehling H, Hiebler S, Ziegler F. Heat transfer enhancement in water when used as PCM in thermal energy storage. Applied Thermal Engineering. 2002;**22**(10):1141-1151. DOI: 10.1016/S1359-4311(02)00035-2

[2] Hawlader MNA, Uddin MS, Zhu HJ. Preparation and evaluation of a novel solar storage material: Microencapsulated paraffin. International Journal of Science Education. 2000;**20**(4):227-238. DOI: 10.1080/01425910008914357

[3] Marinković M, Nikolić R, Savović J, Gadžurić S, Zsigrai I. Thermochromic complex compounds in phase change materials: Possible application in an agricultural greenhouse. Solar Energy Materials and Solar Cells. 1998;**51**(3):401-411. DOI: 10.1016/S0927-0248(97)00259-6

[4] Baby R, Balaji C. Experimental investigations on phase change material based finned heat sinks for electronic equipment cooling. International Journal of Heat and Mass Transfer. 2012;**55**(5-6):1642-1649. DOI: 10.1016/j.ijheatmasstransfer.2011.11.020

[5] Salaün F, Devaux E, Bourbigot S, Rumeau P. Application of contact angle measurement to the manufacture of textiles containing microcapsules. Textile Research Journal. 2009;**79**(13):1202-1212. DOI: 10.1177/0040517508100724

[6] Salaün F, Devaux E, Bourbigot S, Rumeau P. Thermoregulating response of cotton fabric containing microencapsulated phase change materials. Thermochimica Acta. 2010;**506**(1-2):82-93. DOI: 10.1016/j.tca.2010.04.020

[7] Salaün F. Microencapsulation technology for smart textile coatings. In: Hu J, editor. Active Coatings for Smart Textiles. Cambridge: Woodhead Publishing; 2016. pp. 179-220

[8] Zalba B, Marín JM, Cabeza LF, Mehling H. Review on thermal energy storage with phase change: Materials, heat transfer analysis and applications. Applied Thermal Engineering. 2003;**23**(3):251-283. DOI: 10.1016/S1359-4311(02)00192-8

[9] Jahns E. Microencapsulated phase change material. Paper presented at the 4th workshop IEA ECES Benediktbeuern Germany; October 1999. 28-29

[10] Mehling H. Latent heat storage with a PCM-graphite composite material: Experimental results from the first test store. In: 6th Workshop of IEA Annex 10 Phase Change Materials and Chemical Reactions for Thermal Energy Storage; Stockholm. 2000

[11] Narayanan SS, Kardam A, Kumar V, Bhardwaj N, Madhwal D, Shukla P, et al. Development of sunlight-driven eutectic phase change material nanocomposite for applications in solar water heating. Resource-Efficient Technologies. 2017;**3**(3):272-279. DOI: 10.1016/j.reffit.2016.12.004

[12] Vijay Padmaraju SA, Viginesh M, Nallusamy N. Comparitive study of sensible and latent heat storage systems integrated with solar water heating unit. Renewable Energy and Power Quality Journal. 2008;**1**(06):55-60. DOI: 10.24084/repqj06.218

[13] Lu Y, Roskilly AP, Wang Y, Wang L. Study of a novel dual-source chemisorption power generation system using scroll expander. Energy Procedia. 2017;**105**:921-926. DOI: 10.1016/j.egypro.2017.03.417

[14] Fopah Lele A. State-of-Art of thermochemical heat storage systems. 2016. pp. 15-58. DOI: 10.1007/978-3-319-41228-3_2

[15] Regin AF, Solanki SC, Saini JS. Heat transfer characteristics of thermal energy storage system using PCM capsules: A review. Renewable and Sustainable Energy Reviews. 2008;**12**(9):2438-2458. DOI: 10.1016/j.rser.2007.06.009

[16] Lane GA, Shamsundar N. Solar heat storage: Latent heat materials, Vol. I: Background and scientific principles. Journal of Solar Energy Engineering. 1983;**105**(4):467. DOI: 10.1115/1.3266412.

[17] Zeinelabdein R, Omer S, Gan G. Critical review of latent heat storage systems for free cooling in buildings. Renewable and Sustainable Energy Reviews. 2018;**82**:2843-2868. DOI: 10.1016/j.rser.2017.10.046

[18] Chandra D, Ding W, Lynch RA, Tomilinson JJ. Phase transitions in "plastic crystals". Journal of the Less-Common Metals. 1991;**168**(1):159-167. DOI: 10.1016/0022-5088(91)90042-3

[19] Vats G, Vaish R. Phase change materials selection for latent heat thermal energy storage systems (LHTESS): An industrial engineering initiative towards materials science. Advanced Science Focus. 2014;**2**(2): 140-147. DOI: 10.1166/asfo.2014.1091

[20] Abhat A. Low temperature latent heat thermal energy storage: Heat storage materials. Solar Energy. 1983;**30**(4):313-332. DOI: 10.1016/0038-092x(83)90186-x

[21] Kitajgorodskij AI. L'empaquetage de molécules longues. Acta Crystallographica. 1957;**10**:802

[22] Muller A. An X-ray investigation of normal paraffins near their melting points. Proceedings of the Royal Society A: Mathematical, Physical and Engineering Sciences. 1932;**138**(836):514-530. DOI: 10.1098/rspa.1932.0200

[23] Dirand M, Bouroukba M, Chevallier V, Petitjean D, Behar E, Ruffier-Meray V. Normal alkanes, multialkane synthetic model mixtures, and real petroleum waxes: Crystallographic structures, thermodynamic properties, and crystallization. Journal of Chemical & Engineering Data. 2002;**47**(2): 115-143. DOI: 10.1021/je0100084

[24] Mukherjee PK, Dey S. Simple Landau model of the liquid-RII-RI rotator phases of alkanes. Journal of Modern Physics. 2012;**03**(01):80-84. DOI: 10.4236/jmp.2012.31012

[25] Mondal S. Phase change materials for smart textiles—An overview. Applied Thermal Engineering. 2008;**28**(11-12):1536-1550. DOI: 10.1016/j.applthermaleng.2007.08.009

[26] Karkri M, Lachheb M, Gossard D, Ben Nasrallah S, AlMaadeed MA. Improvement of thermal conductivity of paraffin by adding expanded graphite. Journal of Composite Materials. 2015;**50**(19):2589-2601. DOI: 10.1177/0021998315612535

[27] Pokhrel R, González JE, Hight T, Adalsteinsson T. Analysis and design of a paraffin/graphite composite PCM integrated in a thermal storage unit. ASME 2008 2nd International Conference on Energy Sustainability. 2008;**2**:621-629. DOI: 10.1115/es2008-54043

[28] Mondieig D, Rajabalee F, Metivaud V, Oonk HAJ, Cuevas-Diarte MA. n-Alkane binary molecular alloys. Chemistry of Materials. 2004;**16**(5): 786-798. DOI: 10.1021/cm031169p

[29] Kravchenko V. The eutectics and solid solutions of paraffins. Acta Physicochimica URSS. 1946;**21**:335-344

[30] Agafonov IA, Garkushin IK, Miftakhov TT. Patterns of phase diagram changes in series of n-alkane

binary system. Zhurnal Fizicheskoj Khimii. 1999;**73**(5):681-685

[31] He B, Setterwall F. Technical grade paraffin waxes as phase change materials for cool thermal storage and cool storage systems capital cost estimation. Energy Conversion and Management. 2002;**43**(13):1709-1723. DOI: 10.1016/S019689040100005X

[32] Farid MM, Khudhair AM, Razack SAK, Al-Hallaj S. A review on phase change energy storage: Materials and applications. Energy Conversion and Management. 2004;**45**(9-10):1597-1615. DOI: 10.1016/j.enconman.2003.09.015

[33] Sharma A, Tyagi VV, Chen CR, Buddhi D. Review on thermal energy storage with phase change materials and applications. Renewable and Sustainable Energy Reviews. 2009;**13**(2):318-345. DOI: 10.1016/j.rser.2007.10.005

[34] Xie N, Huang Z, Luo Z, Gao X, Fang Y, Zhang Z. Inorganic salt hydrate for thermal energy storage. Applied Sciences. 2017;**7**(12):1317. DOI: 10.3390/app7121317

[35] Dannemand M, Johansen JB, Furbo S. Solidification behavior and thermal conductivity of bulk sodium acetate trihydrate composites with thickening agents and graphite. Solar Energy Materials & Solar Cells. 2016;**145**:287-295. DOI: 10.1016/j.solmat.2015.10.038

[36] De Paola M, Arcuri N, Calabrò V, De Simone M. Thermal and stability investigation of phase change material dispersions for thermal energy storage by T-history and optical methods. Energies. 2017;**10**(3):354. DOI: 10.3390/en10030354

[37] Lv P, Liu C, Rao Z. Review on clay mineral-based form-stable phase change materials: Preparation, characterization and applications. Renewable and Sustainable Energy Reviews. 2017;**68**:707-726. DOI: 10.1016/j.rser.2016.10.014

[38] Sutjahja IM, A U SR, Kurniati N, Pallitine ID, Kurnia D. The role of chemical additives to the phase change process of $CaCl_2 \cdot 6H_2O$ to optimize its performance as latent heat energy storage system. Journal of Physics Conference Series. 2016;**739**:012064. DOI: 10.1088/1742-6596/739/1/012064

[39] Espeau P, Roblès L, Mondieig D, Haget Y, Cuevas-Diarte MA, Oonk HAJ. Mise au point sur le comportement énergétique et cristallographique des n-alcanes. Journal de Chimie Physique et de Physicochimie Biologique. 1996;**93**:1217-1238. DOI: 10.1051/jcp/1996931217

[40] Barbillon P, Schuffenecker L, Dellacherie J, Balesdent D, Dirand M. Variation d'enthalpie subie de 260 K à 340 K par les n-paraffines, comprises entre l'octadécane (n-C_{18}) et l'hexacosane (n-C_{26}). Journal de Chimie Physique et de Physicochimie Biologique. 1991;**88**:91-113. DOI: 10.1051/jcp/1991880091

[41] Metivaud V, Rajabalee F, Oonk HAJ, Mondieig D, Haget Y. Complete determination of the solid (RI)-liquid equilibria of four consecutive n-alkane ternary systems in the range C14H30-C21H44 using only binary data. Canadian Journal of Chemistry. 1999;**77**(3):332-339. DOI: 10.1139/v99-004

[42] Rajabalee F, Métivaud V, Mondieig D, Haget Y, Cuevas-Diarte MA. New insights on the crystalline forms in binary systems of n-alkanes: Characterization of the solid ordered phases in the phase diagram tricosane + pentacosane. Journal of Materials Research. 1999;**14**(06):2644-2654. DOI: 10.1557/jmr.1999.0354

[43] van Miltenburg JC, Oonk HAJ, Metivaud V. Heat capacities and derived thermodynamic functions of n-nonadecane and n-eicosane between 10 K and 390 K. Journal of Chemical &

Engineering Data. 1999;**44**(4):715-720. DOI: 10.1021/je980231+

[44] Salaün F, Devaux E, Bourbigot S, Rumeau P. Development of phase change materials in clothing part I: Formulation of microencapsulated phase change. Textile Research Journal. 2010;**80**(3):195-205. DOI: 10.1177/0040517509093436

[45] Shim H, McCullough EA, Jones BW. Using phase change materials in clothing. Textile Research Journal. 2001;**71**(6):495-502. DOI: 10.1177/004051750107100605